Electrochemistry
for Ecologists

Electrochemistry for Ecologists

J. O'M. Bockris

School of Physical Sciences
The Flinders University of South Australia
Adelaide, Australia

and

Z. Nagy

Diamond Shamrock Corporation
Painesville, Ohio
U.S.A.

PLENUM PRESS • NEW YORK–LONDON

Library of Congress Cataloging in Publication Data

Bockris, John O'M
  Electrochemistry for ecologists.

  Bibliography: p.
  1. Electrochemistry, Industrial. 2. Environmental protection. I. Nagy, Zoltán,
1933-   joint author. II. Title. [DNLM: 1. Air pollution. 2. Ecology. 3. Electrochemistry. 4. Environmental health. 5. Refuse disposal. WA750 B665e 1973]
TP255.B6   621.35   73-84003
ISBN 0-306-30749-9

© 1974 Plenum Press, New York
A Division of Plenum Publishing Corporation
227 West 17th Street, New York, N.Y. 10011

United Kingdom edition published by Plenum Press, London
A Division of Plenum Publishing Company, Ltd.
Davis House (4th Floor), 8 Scrubs Lane, Harlesden, London NW10 6SE, England

Printed in the United States of America

Preface

The present decade might be described as one in which man in the affluent countries is finally realizing that there is a bill to pay at the end of the feast—a feast at which he has eaten without inhibition, without knowing there was any need for inhibition. But now, with the situation fully clear, there is a strange non-plussedness about man's reactions. More oil wells are sought, and clean-up packages are proposed for the same old cars. There is no real awareness yet that this is the end of an era, that quite new technologies have to be built up, and that the time remaining for this is a shatteringly short 15–30 years.

However, there are sources of energy other than the fossil fuels. Oil and natural gas will run out (i.e., become too expensive) in any case during the next two or three decades. There seems little point in waiting until fuel is actually rationed and cars can only be used to move from home through smog to work before we change to these new and clean sources. The time to start the change is now, for there is much to be done.

But new power sources yielding abundant, cheap electricity are no good unless we can *use* the energy cleanly and run an affluent society which does not have a limited lifetime and does not have to be restricted to one-third of the population. We have to put foward proposals to make these things possible, but before we do so, the reality of the "spaceship Earth" concept must sink in: there is just so much material here, and the only way to go on and on using materials is complete recycling.

This then is the background of our attempt to present electrochemical science to the ecologist. It is necessary to state this background to explain the form of the book. We believe we have a big message: that exciting new aspects of a quite old science will be particularly relevant in the newly evolving world we have to build. So, to whom should we address the book? On the one hand, it is clear that the professional working full-time in a scientific field related to energy resources or pollution control is the primary potential reader. On the other hand, the issues are so big and of such general interest that there seems a need to try to communicate to a much, much wider audience, perhaps to the intelligent layman, the *New York Times* reader.

We contemplated producing a journalistic effort, but finally we drew back. Perhaps we should have tried to convince the largest possible number of people of our views, but it is not our aim to produce emotional converts. Our case can be made in terms of facts and ideas, and what we have to say about energy, pollution, and resources is very clear. But understanding it does require *some* science, some chemistry; we want our reader to become convinced by understanding our reasoning, for the penny to drop in his own mind. We want to write for the *Scientific American* reader.

So, finally, we compromised and wrote a fairly chatty main part of the book, which can be understood with high school chemistry. Then we wrote three appendices which can easily be understood by freshman chemists at universities—or indeed by those who know or remember a modern high school chemistry course.

This book is not intended to teach electrochemistry, but to show what a normal spread of information and research in electrochemistry could do to make a balanced world in the future.

One of us (Z.N.) wants to express his appreciation to the management of the Divisional Technical Center (Electrochemical Division) of Diamond Shamrock Chemical Company for their encouragement and help during the writing of this book.

J. O'M. Bockris
School of Physical Sciences
The Flinders University of South Australia
Adelaide, Australia

Z. Nagy
Diamond Shamrock Corporation
Painesville, Ohio 44077
United States of America

Reading List

R.W. Murray and C.N. Reilly, *Electroanalytical Principles,* Wiley, New York, 1963.

B.E. Conway, *Theory and Principles of Electrode Processes,* Ronald Press, New York, 1965.

R. Jasinski, *High-Energy Batteries,* Plenum Press, New York, 1967.

J.O'M. Bockris and S. Srinivasan, *Fuel Cells: Their Electrochemistry,* McGraw–Hill, New York, 1969.

J.O'M Bockris and A.K.N. Reddy, *Modern Electrochemistry,* Plenum Press, New York, 1970.

A.T. Kuhn (editor), *Industrial Electrochemical Processes,* Elsevier, New York, 1971.

J.O'M. Bockris and D. Drazic, *Electrochemical Science,* Taylor and Francis, London, 1972.

J.O'M. Bockris (editor), *Electrochemistry of Cleaner Environments,* Plenum Press, 1972.

L. Antropov, *Theoretical Electrochemistry,* Mir Publishers, Moscow, 1972.

J.O'M. Bockris and R. Fredlein, *Workbook of Electrochemistry,* Plenum Press, 1973.

J.O'M. Bockris, N. Bonciocat, and F. Gutmann, *A Primer in Electrochemistry,* Wykeham Press, London, 1973.

Contents

Chapter 2
Some Consequences of the Present Energy Policy

Chapter 3
Future Energy Sources

Chapter 4
Electrochemical Sources of Power: Batteries

Chapter 5
Electrochemical Sources of Power: Fuel Cells

Chapter 6
The Hydrogen Economy

Chapter 7
Electrochemical Waste Treatment

Chapter 8
Electrochemical Methods of Waste Disposal

Appendix II
Interfacial Charge Transfer

Appendix III
Transport of Charges to and from an Interface

CHAPTER 1

Ecology and Electrochemistry

1.1. The Rapid Development of Fears about Pollution

In November of 1971, twenty-three large industrial plants were ordered to cease operation in Birmingham, Alabama, by a U.S. Court because of imminent health hazard to the population of the area due to air pollution. In major cities, an "air pollution index" is routinely given in the weather report just like the temperature. The daily papers devote considerable space to ecologically oriented news and columns. This attention given to pollution and the future of man is a sudden thing. In the early sixties, air pollution was considered a peculiarity of Los Angeles, and newspapers were devoid of references to ecology. By 1970, they were full of them. The problem is not likely to leave us. The pollution of our waterways has occurred before our eyes to such an extent that the collected trash and waste on and in them constitute a fire hazard. In the summer of 1969, a river (the Cuyahoga in Ohio) actually did catch fire.

Our rivers, our lakes, even our seas, are undergoing a deterioration which is dramatically rapid. The dead fish in the lakes are a vivid reminder of the relation of one aspect of an eco-system to another.

Why have these ecological matters come so suddenly to the public's attention? One reason is the fast accumulation of effects.

People can see disaster in their own lifetime, not in some vague future, when only other generations can be affected.

Another reason why pollutional anxiety has come upon us arises from the fact that the energy use of man has undergone a great upthrust during the last few years (Fig. 1.1). Energy use per person and the standard of life probably run parallel (*cf.* the relation between the energy consumption and the gross national product of nations, shown in Fig. 1.2). The suddenness of the terror comes from the fact that our standard of life has increased greatly, just in the last few decades. We did not realize that it would give us a tremendous hidden bill to pay, the necessity of making a radical change in our system of obtaining energy and dealing with our waste products.

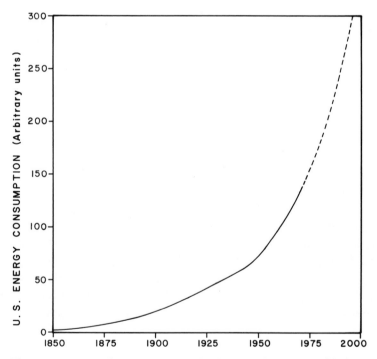

Fig. 1.1. Increase of energy consumption in the United States with time.

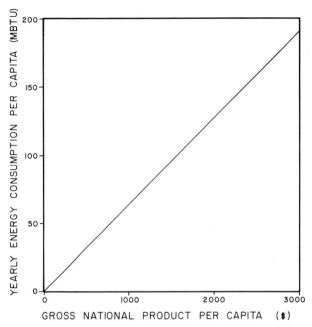

Fig. 1.2. Relation between the gross national product and
the energy consumption of different countries.

1.2. The Most Pressing Ecological Danger:
Damage to the Atmosphere

When man was not industrial, had not started to dig up
the green plants for building space, and coal and oil for burning—
converting them to carbon dioxide—there was an equilibrium
between the CO_2 produced by metabolism, the decay of dead
matter, and its evolution from volcanoes, and the consumption
of carbon dioxide in photosynthesis to form plants, with the
exudation of oxygen into the biosphere. This is a mobile equilibrium and, for geological reasons, the amount of carbon dioxide
in the atmosphere has gone through a number of cyclical
variations during the ages. In general, these changes of the
carbon dioxide content of the atmosphere can be reasonably
well connected with climatic changes in the past; the increase of

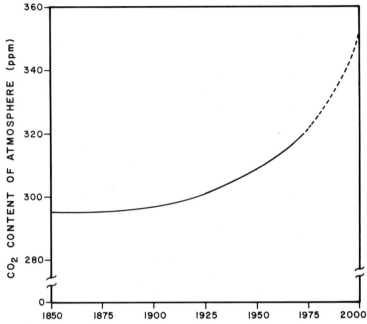

Fig. 1.3. Accumulation of carbon dioxide in the atmosphere.

CO_2 content is associated with warm periods, and its decrease with glacial periods.

It is only since around the turn of the century that man has become a significant factor in this equilibrium. Paralleling the increase in power consumption, man's activities resulted in a considerable increase in the carbon dioxide content of the atmosphere, as shown in Fig. 1.3. It can be predicted that, if the use of fossil fuels to produce energy continues along its present trend, and if there are no counteracting factors, a significant ($> 1°C$) rise in the temperature of the earth will result by *ca.* 2000 A.D. There is a corollary: the melting of ice in areas such as Greenland and the southern polar continent. The added water will cause the sea level to rise, and in some regions there may be, even by 2000 A.D., flooding. A significant loss of available territory seems likely by 2050 A.D. if we continue to eject carbon dioxide into the air along an extrapolation of the present line of CO_2 *versus* time.

1.3. The Necessity to Stop the Burning of Fossil Fuels as the Source of Energy at the Earliest Moment

At the present time, there is little consciousness among the public *that it is the burning of fossil fuels*, with the atmospheric and climatic changes to which this is bound to give rise, which is the chief menace from pollution. It is what they see which impresses them most, and they see, e.g., detergent suds in the river, and dead fish. However, a greater long-range danger is the pollution of the atmosphere by carbon dioxide (accompanied by other more well-known pollutants, e.g., CO, unsaturated (smog-causing) hydrocarbons, NO_x, etc.), which is caused by the skyrocketing increase in the consumption of energy obtained from fossil fuels. The fact that the overall efficiency in the energy conversion with present methods is less than 30% further aggravates the situation.

The environmental scientist is aghast when he hears excitement over Alaskan oil, Australia's discovery of natural gas, or some new internal combustion engine which has a reduced volume but pollutes like all the others. He knows that there is a very limited time in which a continuation of the old economy is possible. He would rather see the excitement go to the evolution of new methods of providing man with the energy he needs, and which will last him not another decade but for the foreseeable future. The first step in this direction is the elimination of the burning of oil, natural gas, and above all coal (because its burning ejects large quantities of SO_2 into the atmosphere), and the introduction of alternative energy sources.

1.4. Replacement of Fossil Fuels

What kind of energy sources will replace fossil fuels? Atomic technology promises to be one new source. It is not without its own difficulties, however, because the pollution it causes, though much less in amount (and also much less visible) than that due to fossil fuels, has its dangers nonetheless. The use of the sun's radiation, geothermal sources, and tidal (gravitational) sources have been investigated in the past. Abundant energy from the sun, effective for evermore, is available, but transduction of this energy into electric power has not yet proved economical. It is very largely a matter of research. That devoted to the use of atomic energy amounts to several times more than that devoted to all the others energy sources put together. With the realization of the imminent pollution hazards inherent in the further use of fossil fuels, the economics of the alternative sources could be taxed into desirability.

With the approaching exhaustion of the fossil fuels, there will in any case be price rises which will make research on new sources socially desirable, and hence attractive for the action of politicians.

The new sources of energy will be discussed in detail in a later chapter. Here only some common characteristics will be emphasized. They will have to be located far from the user of the energy—the atomic units for safety's sake, and because the heat evolved by large atomic reactors would make their placement near towns unacceptable. Solar sources would be placed on the sea in the southern hemisphere or in the open spaces of Australia, Africa, or India. They gain greatly in efficiency if built in large size. None of the new sources is suitable for transportation use—they cannot be made to fit under the hood of a car. The energy produced by the new and remote sources will have to be moved around in a medium which can be easily transported and distributed. Presently, energy is transported in the form of gas or oil to the generating plant, or as electricity through wires. What will the transport mode for the energy of the future be?

1.5. Electricity and Hydrogen: The Only Media of Energy

As oil and natural gas become unavailable, electricity will become increasingly the predominant form in which energy will be used, but it is expensive to transport over long distances. There is a voltage drop through the wires due to their electrical resistivity. This is proportional to distance, and because of this the cost of transportation for distances over 500 miles makes up some 50% of the cost of the electricity at the delivery point. A further, environmental consideration is the increasing despoilation of the country by overhead wires, especially if one thinks of the prospect that, with exponentially increasing power requirements, the number of wires crisscrossing the country will increase exponentially. But, if the cables are laid underground, there can be an increase in the cost of the transportation of energy up to ten times.

The idea suggests itself that electricity might be used at the point of generation to produce a "synthetic fuel," which can be transported more economically than electricity and transduced to energy at the point of use. Hydrogen is the first contender. It is easy to produce by the electrolysis of water, can be stored easily in gaseous and liquid form, is easy to transport through pipelines, and can be converted back to electricity at the location of the consumer (plant or private residence) with fuel cells (*cf.* the successful Gemini and Apollo fuel cell systems of the space program). It can also be burned to produce heat directly. It can be used directly as a transportation fuel, by combusting it in engines similar to the present ones (the only product would be pure water), or used to drive cars silently and cheaply by converting it to electricity in on-board fuel cells, driving electric motors.

1.6. Implications for Electrochemical Science

The fact that the media of energy in the near future will be electricity and, probably, hydrogen has implications for the use of electrochemical devices on a wide scale. They become essential in the concept of the hydrogen economy because of the advantages of electrolyzing water to hydrogen and oxygen and using hydrogen–air fuel cells to produce electricity at the site. However, electrochemical methods become important in other places too, partly because the availability of low-cost electricity will make its direct use in technology favorable, and partly because there won't be another fuel to use. Furthermore, electrochemical methods in technology allow better control of chemical reactions, having a third lever, the applied voltage, for precise control, in addition to the classical ones of temperature and pressure. For example, electrochemical methods are now used in the deposition of metals from solutions containing metal salts. As electricity becomes cheaper—and the only available energy source—its applications in chemistry, chemical engineering, and metallurgy will become more numerous. The synthesis of organic compounds will increase, and detoxification of factory wastes, selective recovery of valuable metals from liquids, regeneration of factory-used solutions, etc., will all be carried out electrochemically. A difference between such electrochemical methods and the normal chemical methods now in use is that the electrochemical ones can be more easily made to avoid the exudation of waste products into the atmosphere.

1.7. What Is Electrochemical Science?

It is easy to answer this question: it is the interdisciplinary activity which has grown up around phenomena connected with charge transfer at interfaces (Fig. 1.4).

Modern electrochemistry was derived from an older, more diffuse, idea. The term formerly meant the physical chemistry of solutions—for example, pH, ionic equilibria, the conductivity of electrolytes, thermodynamic measurements with cells, etc. But, more recently, it has been refined to mean: dealing with, in, and around phenomena arising from electron transfer at interfaces.

As will become clear in the reading of this book, there is electrochemistry in most things connected with surfaces in contact with liquids. The illustrations used will deal with metal electrodes dipped in aqueous solutions, where charge transfer occurs at the metal-solution interface as current is passing through the system resulting in a desired chemical change (or a chemical reaction is producing electrical current). The subject

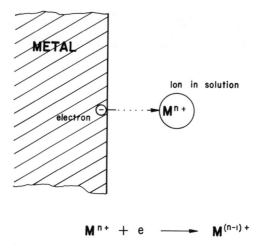

$$M^{n+} + e \longrightarrow M^{(n-1)+}$$

Fig. 1.4. Simple picture of a charge transfer. An electron from the metal transfers to an ion in the solution, changing the oxidation state (charge) of the ion. This is half a chemical reaction.

is broader, however, for it also applies, e.g., when the "liquid" is an invisible moisture film on metals, as it sometimes is in the case of corrosion (see Section 9.7). Further, the electrode does not have to be a metal but can be a semiconductor or an insulator. The "liquid" does not have to be an aqueous solution but can be an organic system, a molten salt, or an ionically conducting solid. Electrochemistry has further-out connections, for example, to biophysics and biology; it is known that a change in the electrical potential across the nerve cell-solution interface is an essential part in their functioning. And, since the brain exhibits' continuous oscillatory electrical activity, the origin of which cannot be electromagnetic, it may not be too far-fetched to regard the brain's activity as that of an electrochemically operating computer. Some detailed suggestions in this direction have recently been made.

1.8. The Coming Change to Electricity and Hydrogen as the Media of Energy Implies an Extensive Electrochemical Technology

Clearly, if the medium of energy is electricity, and its origin solar or atomic, there will be little *chemical* burning of fuels as the origin of energy, and the fume-emitting chemical technology of the present will have a diminishing technological role. Chemical technology will become a matter of the 19th and 20th centuries. Electrochemical technology will tend to become the major technology in the 21st century. Electrochemical technology can make most of the things which man desires possible, with zero ejection of pollutants into the atmosphere. Correspondingly, it provides the basis of small and mobile power sources, applicable in moving vehicles; allows a clean metallurgy, including the recycling of metals; a sewage system which can be made to give end products of high energy, e.g., methyl alcohol; and becomes the basis of clean energy distribution.

1.9. Purposes and Limitations of This Book

This book is intended to give the ecologically-minded reader a knowledge of some electrochemistry so that he can understand the devices to which electrochemistry does and could give rise, and which could replace many of the present polluting devices and industrial processes.

Study of electrochemistry included in this book is not sufficient to make the reader an electrochemical engineer. However, it is sufficient to give to an ecologically oriented reader enough information of the principles and possibilities of the field to enable him to think electrochemically for the future. It is a trigger for specialist studies, and an information source for the man who wants to know one of the long-term transformations along which a postindustrial society will run.

There are two things which this book will not do. It will not retrain an engineer so that he can work independently as an electrochemical engineer. It will not present full electrochemical solutions to any of the problems, but only the outline of the principles by which a solution seems possible if the principles enunciated were developed into engineering processes.

1.10. Research and Development in Electrochemistry

The research and development in electrochemical science needed to increase our knowledge of the field sufficiently to to make its principles practical is vast. The principles of modern electrochemistry were built up during the 1950's. It was during the 1960's that the interdisciplinary aspects of the new electrochemistry were realized, and only in 1970 that the first comprehensive book on modern electrochemistry was published. It is a new field, largely awaiting development. It is particularly appropriate to ecological purposes in that it emphasizes cyclic recovery of previously used materials. And it would lead to the end of air and water pollution, i.e., a reduction of disturbance of the naturally balanced condition.

But there is a negative side to electrochemical science. Few people know what it is. Many people, particularly chemists, think of a subject different from that presented in this book. They think of thermodynamic electrochemistry, electroanalytical chemistry, or perhaps the older electrochemistry of solutions. We are talking about the electrochemistry of solid surfaces, electrochemical science, the interdisciplinarily-oriented applied electrochemistry which is a subject synthesized during the last 20 years by the application of solid state physics to considerations of the interface between solids and liquids (see Table 1.1).

But there is another difficulty about electrochemistry, one which will not yield to changes in educational policies nor, immediately, to increased government funding of research. It is that a number of concepts developed in this book are at the present time scientific ideas which have often been tested in laboratories but for the most part have not yet been engineered in practice. They are sound ideas, and would not have been put in the book if they had not been subject to the best numerical

TABLE 1.1   Connections of Electrochemistry with Fundamental and Applied Sciences

Fundamental Sciences	Electrochemical Science	Applied Sciences
Kinetics (charge transfer kinetics at interfaces)	Electrodics (the structure of electric double layer, the kinetics and mechanism of charge transfer reactions at interfaces)	Energy storage (batteries, electrically driven automobiles, storage of atomically derived electricity)
Quantum mechanics of charge transfer processes		Energy conversion (fuel cells, conventional fuels or synthetic fuels derived from atomic energy, dry cells)
Theories of aqueous solutions (transport properties, diffusion, conductivity, structure)		Pollution abatement (electrochemical waste destruction, electrodialysis, detecting devices)
Theories of ionic liquids (molten salt structure, transport properties)		Metallurgy (new processes with nonpolluting electrochemical hydrometallurgy, electrowinning, electrorefining, aqueous or molten salt systems)
Surface science (double layer structure at solid-liquid interfaces, colloid systems)		Metal fabrication (electrochemical machining, electroforming)
Thermodynamics (determination of thermodynamic parameters)		

Catalysis (electrocatalysis at electrodes, general catalysis)

Inorganic chemistry (preparation technique)

Organic chemistry (preparation technique)

Analytical chemistry (polarography, potentiometry, amperometry, coulometry)

Crystallography (electrocrystallization, morphology)

Biology, biophysics (current in nerves, membrane equilibria)

Ionics (the structure and properties of solutions and ionic liquids)

Corrosion protection (inhibitors, cathodic and anodic protection, stress-corrosion cracking)

Material recycling (recovery of metals from waste materials including junked cars)

Surface treatments (electrocoating, electroplating, anodizing)

Water desalination (electrodialysis, water production from fuel cells in the hydrogen economy)

Production of chemicals (adiponitrile, chlorine, fluorine, sodium hydroxide, sodium chlorate, hydrogen, oxygen, heavy water)

Biology, Biophysics (nerves, thrombitic, digestive, bone growth) estimates of practicality possible without further research. But the ecologist will be the first person to realize that something calulated with parameters correct for an isolated situation may have to be modified when introduced into a *system*. Thus, in electrochemistry, an enormous amount has still to be *developed* and *engineered*, taken from the fundamental laboratory to the chemical engineering lab. The people who will have to do all this in the next two or three decades ought to know something about the field before they start to work on it. Due to the lack of teaching in universities in America in the new electrochemistry, they have not in the recent past known the basics of the systems considered. It is, then, *electrochemical systems engineering* which is a subject of great importance for the future, the realization in a practical sense of what the vistas opened up by the new electrochemistry in the last 20 years have shown to be possible.

CHAPTER 2

Some Consequences of the Present Energy Policy

2.1. The Short-Range Effect of Burning Fossil Fuels

The continued burning of the fossil fuels to obtain our energy needs has several effects, each appearing on a different time scale. The one threatening us in the shortest time is understandably the most well known. This is the fouling of the atmosphere with smog-forming materials which is causing an immediate and increasing health hazard for the population in most industrial towns. Such effects are occurring wherever there are big cities. The basis of such effects is the incomplete burning of the fossil fuel, resulting in carbon monoxide and unsaturated hydrocarbons. Also, other side products of the energy-producing reaction, like nitrogen oxides, sulfide dioxide, and soot, are present. The main polluters are our oil- and coal-burning generating plants and cars. Pollution arising from these sources has been newspaper material since the late 1960's.

Some action is being taken to clean up these pollution sources. However, will the partial cleanup be a long-range solution? It seems unlikely that it will be possible to use the internal combustion engine for much longer, particularly so long as the number of cars in the world continues to increase. That it will do so seems likely in most countries, except in the most developed—the U.S.A. and U.K.—where a tendency toward saturation of roads, and perhaps needs, will become apparent during the next one or two decades. It can be shown that, world-wide, the present rate of increase in the number of cars will overcome any reasonably expected improvement in the average rate of pollution per car before the end of the century. There-after, the rate of atmospheric pollution, which is expected to decline during the next 15–20 years because of the reduction of pollution from each car, will again increase. Similar reasoning can be applied to coal- and oil-burning electricity generating plants, because of the exponential rise of electricity needs associated with generally increased living standards.

2.2. An Effect of Burning Fuels to Their End

In the somewhat longer run, we have to face the exhaustion of our fuel resources. As far as oil and natural gas are concerned, the proven reserves will be exhausted in the United States before 1985, and were we to use oil thereafter, it would all have to be imported mostly from the Middle East (where it will exhaust about 2000 A.D.). Were we to burn coal instead, the pollution would be likely to be worse than that obtained by burning oil. By continuing to burn our oil and natural gas to the point of exhaustion, we are losing precious raw materials for our future chemical industry, especially synthetic textiles and eventually synthetic proteins. But such synthetic materials become more and more important as time goes on (for the population growth tends to outgrow the former supplies of these materials).

2.3. A Little-Known Effect of Fossil Fuel Burning

There is much less awareness of another long-range effect, namely, that of the increasing CO_2 content in the atmosphere on the climate. This effect will make it highly undesirable, and eventually impossible, to use fossil fuels in central power stations or in transportation, no matter what means might be developed for eliminating unsaturated hydrocarbons, CO, SO_2, and NO_x, from the result of burning fossil fuels.

Carbon dioxide is the main product (*ca.* 98%) of the reactions of fossil fuel combustion. It will continue to be emitted to the atmosphere regardless of the efficacy which might conceivably be developed for the removal of the other substances.

The build-up of this pollutant to a concentration at which it affects the climate is one of the stronger reasons for the transfer to an electrically operated technology and the complete elimination of the burning of fossil fuels. Extrapolation of the present rate of increase of CO_2 concentration with time indicates that significant effects must be expected by 2000 A.D.

2.4. The Temperature of the Atmosphere Depends on the Carbon Dioxide Content

The principal constituents of the atmosphere, oxygen and nitrogen, do not absorb electromagnetic radiation. However, the three secondary constituents (carbon dioxide, water vapor, and ozone) do absorb it, mainly in the infrared and ultraviolet regions.

The temperature at the surface of the earth is a balance between heat from the internal radioactivity of the earth, heat from the sun, the loss of heat to space, and reflection of a fraction of the sun's light back through the atmosphere to space. Carbon dioxide in the atmosphere absorbs some of the reflected solar radiation in the infrared part of its spectrum, and degrades it to translational energy, i.e., heat. It is this heat which will tend to cause the increase in temperature. An effect of this kind is sometimes termed "the greenhouse effect."

The CO_2 theory of the climate was first proposed by Tyndall in 1861. Extensive work has been needed to establish its validity. The leading names in this area are those of Plass and, particularly recently, Manabe.

Calculations concerning the relation between the concentration of carbon dioxide in the atmosphere and its temperature rise are complex, for they involve knowing precisely the structure of the various absorption bands of carbon dioxide, and then taking into effect a number of complicating factors. Some of these are:

i. *The Effect of Water Vapor*. This is a reinforcing effect. Water in the atmosphere causes a temperature effect similar to that of CO_2. Therefore, if the atmosphere's temperature tends to rise because of the increase of CO_2, increased water evaporation will cause a further temperature rise.

ii. *Changes in the Albedo*. The efficiency of reflection from the earth of light (for example, the reflectivity of the polar ice) can be lessened if pollutants from the atmosphere—dust and solids—condense on the reflecting surface. This could result in an increased absorption of solar light, and a consequent increase of temperature greater than that predicted by Tyndall's CO_2 theory.

iii. *Clouds*. These clearly affect the earth's temperature. Estimates are usually made for average cloudiness. Feedback effects are difficult to calculate because of the lack of precise knowledge of the effect of temperature changes upon the cloud concentration.

iv. *The Amount of Particulate Matter in the Atmosphere*. This acts counter to the temperature-increasing effects of CO_2 and decrease in albedo.

The best estimate, which takes into account every one of the presently calculable factors, predicts a temperature rise of about 2.5°C (4.4°F) by a doubling of the carbon dioxide content of the atmosphere.

2.5. The Carbon Dioxide Balance

A number of factors affect the carbon dioxide balance, and Table 2.1 (due to Plass) summarizes these. Thus, the amount of CO_2 contributed to the atmosphere by burning fossil fuels is considerable (*about 1/7 of the amount withdrawn by plants on land in photosynthesis*). It represents an effect with which photosynthesis did not have to deal prior to this century. It has upset

Table 2.1. Factors Affecting the Carbon Dioxide Balance

	Exchange activity (in units of 10^9 metric tons of carbon per year)
Withdrawn from atmosphere in photosynthesis by plants on land	35
Returned to atmosphere in respiration by plants on land	10
Plants and animals become dead organic matter on land	25
Released to atmosphere by decay of dead organic matter on land	25
Withdrawn from atmosphere by exchange with surface layers of ocean	100
Returned to atmosphere by exchange with surface layers of ocean	97
Withdrawn from surface layers of ocean in photosynthesis by plants in ocean	40
Plants and fish become dead organic matter in ocean	40
Released to surface layers of ocean by decay of dead organic matter in ocean	35
Released to deep oceanic layers by decay of dead organic matter in ocean	5
Withdrawn from deep ocean and added to surface layers by upwelling water	45
Withdrawn from surface layers and added to deep ocean by descending currents	40
Dead organic matter forming sediments	<1
Returned to atmosphere by combustion of fossil fuel	5
Removed from atmosphere by weathering of rocks	0.05
Returned to atmosphere by volcanic emissions and hot springs	0.05

the balance between CO_2 production and its removal—hence the increase experimentally observed since about 1900.

The experimentally observed rise in carbon dioxide in the atmosphere and a plot of the extra amounts injected from the burning of fossil fuels are parallel, suggesting that the observed CO_2 increase in this century is indeed due to our use of fossil fuels. At present, we have about 320 ppm of carbon dioxide in the atmosphere and, because of our use of fossil fuels to obtain energy, this concentration is presently increasing at a rate of about 1 ppm per year.* The number of tons of CO_2 being added by fossil fuels is about 10^{10} tons per year and, if we continue to burn fossil fuels to obtain energy, may be as high as 5×10^{10} by 2000 A.D.

2.6. The Effect of the Sea on the Carbon Dioxide Equilibrium

The sea could, if given sufficient time, absorb the excess carbon dioxide which we put into the atmosphere, and thereby cut down the rising amount.

However, the taking up of equilibrium between CO_2 in the atmosphere and the sea is very *slow* indeed. Firstly, the carbon dioxide does not simply dissolve in the sea, but undergoes a series of complex equilibrations, forming carbonates and bicarbonates with the calcium, magnesium, etc., contained in sea water. Secondly, while equilibration with *surface* water is relatively fast, an exchange of the sea water from the surface to the deep oceans has been estimated to last one to ten thousand years. A similar time span may be expected for atmospheric carbon dioxide to reach equilibrium with the bulk of the oceans.

Thus, we cannot expect, in any time of interest to us with respect to CO_2 effects on the atmosphere, significant help from the sea in restoring the imbalance of the CO_2 content of the atmosphere which is caused by the use of the burning of coal, oil, and natural gas to obtain energy.

2.7. Effect of Removal of Photosynthesizers

One of the sources of the removal of carbon dioxide (see Table 2.1), is the photosynthetic reaction of CO_2 with plants. This, however, is a diminishing resource for two reasons. The spread of civilization gives rise to a greater degree of the clearing

---

* The average expected rate of increase per year between 1950 and 2050, using an extrapolation based on the 1900–1950 graph, is about 4 ppm per year.

TABLE 2.2. Carbon Dioxide Added to Atmosphere by Consumption of Fossil Fuels, and the Resulting Temperature Rise

Decade	Atmospheric concentration, ppm	Predicted temperature rise, °C
1860–69	295	0.00
1870–79	296	0.00
1880–89	296	0.01
1890–99	297	0.02
1900–09	298	0.03
1910–19	300	0.04
1920–29	302	0.06
1930–39	305	0.08
1940–49	307	0.10
1950–59	312	0.13
1960–69	317	0.18
1970–79	325	0.24
1980–89	336	0.33
1990–99	352	0.46
2000–09	373	0.63
2010–19	405	0.88
2020–29	449	1.2
2030–39	510	1.7
2040–49	597	2.4
2050–59	718	3.4
2060–69	889	4.8
2070–79	1130	6.7

away of green material every year. Thus, agriculturally-used land is less active in removing CO_2 by photosynthesis than is a wild forest. Correspondingly, the sea-borne plankton (which carry out more photosynthetic removal of carbon dioxide than do the plants on land) may be expected to be increasingly affected by oil spills and other ocean pollution.

2.8. Predicted Increase of Carbon Dioxide

The most recent calculations of the addition of carbon dioxide to the atmosphere and the predicted temperature rise are those of Plass. His values are given in Table 2.2. These calculations were made on the assumptions that about 40% of the carbon dioxide injected into the atmosphere by the burning of fossil fuels stays in the atmosphere and 60% goes into the sea (though it is not in equilibrium with it for the very long time periods mentioned). The effect of a possible decrease in the plankton concentration is not accounted for in this calculation.

2.9. Results of the Increase of Temperature of the Atmosphere

One possible effect of increasing atmospheric CO_2 would be an increase in the average sea level throughout the world. Thus, there are certain areas (Greenland, South Polar Region) where the land is covered with a thick ice sheet. That part of this ice which is now at about $-1°$ C would melt if the average atmospheric temperature at sea level increased by $1°C$, and there would be an increase in the content of the seas.

Very approximate calculations suggest that there would be a rise in sea level of about 25 cm by 2000 A.D., caused by the predicted rise of slightly more than $\frac{1}{2}°C$ in temperature. This rise of nearly one foot would not be very dangerous. However, should we continue to use coal to power our technology, after the oil and natural gas are exhausted, and if world technology continues to expand along the same line it has followed for the last seven decades, then by 2070 A.D. (three generations from now), the world temperature would rise about $7°C$. Such a rise would cause a world sea level rise of some seven meters (*ca.* 20 ft).

2.10. Possible Delaying Tactics to the Effect of Atmospheric Pollutants

Are there ways by which these city-destroying trends could be delayed, while we burn up our coal after the oil and natural gas have all gone? An increase of the efficiency of our energy conversion techniques would be a help. For example, automobile engines have efficiencies below 25%, which means that one has to emit three times as much CO_2 than that which corresponds to the energy gained. Central electricity generating plants are not better than 40% efficient. No improvement is expected, because their efficiences are Carnot limited, and partly because the technologies concerned have reached a point where their further improvement in the direction of higher efficiency would cost more than the gains by the resulting efficiency increase.

The electrochemical conversion of the energy of fossil fuels (*cf.* Chapter 5) would provide temporary relief, since it can work at some 60–80% efficiency. This could buy us perhaps an extra decade or two before the CO_2 atmospheric pollution forces us to abandon the burning of fossil fuels.

The removal of carbon dioxide from effluents, by chemical means, could be considered, but we are talking about 10^{10} tons per year. Whatever product we made from the CO_2 would then become a major world pollutant. The only solution is to stop emitting CO_2, i.e., to stop burning fossil fuels.

It is to be stressed that there is little uncertainty about the increase in CO_2 in the atmosphere so long as fossil fuels continue to be burned as the main path to energy. However, the statements made concerning the resulting temperature, and the consequent sea level effects, are subject to the caveat that some other effect of fossil fuel burning might intervene in a different direction and perhaps reduce the effects predicted and even reverse them. One such possibility is already sighted: the increased particulate matter to which fossil fuel burning leads. The presence of an increasingly significant world pall—essentially a dilute aerosol layer—would *decrease* the heat reaching the earth. Some calculations which stress *this* effect have predicted that it would compensate the temperature rise which the CO_2 is causing, and eventually lead toward a substantial world temperature decrease. Which of the two effects will predominate at what time isn't clear right now: but it *is* clear that burning fossil fuels will have some significant effect on the atmosphere's temperature, and that such an effect will be appreciable by the end of the present century.

2.11. What Is the Present Energy Policy in the U.S.A.?

In recent years the energy policy of the U.S.A. has had a low profile, and has not been enunciated clearly. What it appears to be is a slow advance toward an increase in atomic power by breeder reactors. According to present plans, 26% of electric power *should* be atomic by 2000 A.D. However, prediction of such a percentage neglects the delaying tactics of local administrations, arising from the (partly rational) fears of air pollution from atomic plants.

Even were the present plans to be fulfilled, they would be out of step with the exhaustion of U.S. oil and natural gas, which calls for another energy source to be *effective* before 1990. Research and development for such a source would depend on vast government investment and organization. With a NASA-like setup, perhaps 10 years of research could produce the basis, and it is difficult to see less than a further 20 years for the necessary period of power station building and changeover.

But suppose the vast change to electric power from breeder reactors were to be brought about in time. Then, one would be met with two dilemmas. First, there would be an element of chagrin that all had been rebuilt only to face a pollution situation different (no smog or stink, no climatic change) but possibly worse (increased number of deaths from leukemia, and possible genetic defects caused by atmospheric atomic pollution). Second, atomic reactors produce electricity at central power stations. With what fuel should we run the transportation system?

2.12. Internal Political Aspects of the Relation of Pollution to an Energy Policy

Once upon a time, scientists and engineers believed that if they solved a problem, the solution would be used. This was all the more so in the dynamic capitalistic Western societies, because capitalistic drive made anything which the public wanted, or could be stimulated to want, a ready source to great profit for the entrepreneur, who was then galvanized into activity to provide the need.

However, this situation has been greatly modified by the conversion of the eager, individualistic entrepreneur into giant companies, the presidents of which sometimes control a greater sum of money than the governmental budgets of medium-sized nations. One of the effects of these giant companies is inertial. Once they started upon the manufacture and sale of a product (say, aluminum) on a large scale, they must invest very large sums to build a certain kind of plant in a certain way. Into the calculation of the profitability of the enterprise goes the amortization of the capital used in the building of the plant. It is assumed that the plant will last, e.g., for 15 years. A breakdown of this assumption—caused perhaps by the advent of a new technology "too soon"—will upset profitability. The concepts involved are no more Machiavellian than those of a house owner who saw his house bulldozed away before he had paid off the mortgage. But they do have greater societal impact.

In 19th century capitalism, a new entrepreneur, using the new technology, would come forth to sell the better—or cheaper—product; the public would gain—and the entrepreneur stuck with the old technology undergoing amortization would go under. In 20th century capitalism, the giant corporation which is threatened with a significant change in technology is not likely to be stricken. It can buy the new technology, patent it, but need not use it. It can do a great deal of defensive research on it and advertise the negative side—there always is one. If public opinion becomes a factor, it can be educated by suitable use of the media. If government committees threaten legislation to force a change, the company can make a feint in the direction of change—thus demonstrating the lack of necessity for such legislation—or use the power of lobbying to protect their interests. The old technology will continue to be used and the profits made.

Since the Second World War, the power of the socialist countries has been significant. It is a characteristic of these countries that they invest a larger percentage of their gross national product in education and science than do the Western democracies. These countries should not be put off from the introduction of a new technology by capitalistic considerations. We may thus see them solve their pollution problems before we do!

This is possible, and is of course the public stance of socialist regimes. However, one of the stronger planks of the capitalist case is very true: enterprise and dynamics *are* much greater under a capitalist system, with startling possibilities for the aggregation of personal wealth for the very energetic, in contrast to systems where enterprise and drive are rewarded only by symbolic awards. Less obvious is the fact that the socialist countries, though very strong in fundamental scientific research, are weaker in applied research where the lack of the personal monetary incentive has a soporific effect.

It is easy to see the applicability of these thoughts to energy policy and pollution. Were the policy in the U.S.A. public oriented, NASA would be rapidly followed (and perhaps replaced) by NERDA (the National Energy Research and Development Administration), research on atomic fusion greatly increased, and, above all, investment made in solar energy conversion and a clean electrochemical technology. However, the present energy industry will oppose this with their enormous weight, and overwhelming (of course, latent) political power. On the other hand, just because they constitute a large part of the total economy, any suddenly enforced change in the automobile and energy industries might temporarily hurt the national economy. In this respect, their delaying influence could be justified. However, the time is short and we must not sacrifice our future for our present. A reasonable compromise is needed. (See also Section 5.6.iv.)

2.13. External Political Consequences of the Present Energy Policy

The oil and natural gas supply of the United States will be exhausted in about one decade, and during this time imports of foreign oil will grow toward 100%, the cost to the United States being about $30 billion per year of foreign exchange.* The negative economic effects are considerable. The countries upon which the United States will then have to rely absolutely will be largely the Arab states. Formerly, such states had no countervailing power, so that a too precipitous increase in price, or outright refusal to continue to sell to foreign companies, brought the implicit threat of invasion. Today, however, small nations can turn to the Soviet Union for their weapons and supplies, or even obtain a Russian military guarantee.

Between the early 1980's and the time (no date at present even envisaged) at which the United States will become independent of imported oil, it seems likely that very considerable increases of the price of energy will be enforced.

2.14. Conclusions

Before 2000 A.D. (when one of the two climatic effects of continued fossil fuel burning will become significant), we must reduce our use of fossil fuels to obtain energy, not only in the running of transports, but in all technology. The increase of health hazards due to pollutants which will occur by the end of the century, in spite of pollution abatement, because of our skyrocketing use of energy is an added reason. And so is the necessity to conserve our carbonaceous raw materials for future (non-energy) needs.

According to present plans, the United States will still get 74% of its energy from coal, oil, and natural gas in 2000 A.D. Thus, an acceleration in plans for building atomic reactors, and a tremendous expansion in developmental research on the economic collection and transmission of solar energy, seems needed.

The major impediments to action and attainment reside in internal U.S. politics. It is not easy to fit the profit motive and development of atomic and solar energy into the capitalist system. The main countervailing point is that the inactive policy appears to head toward a torturing situation in external politics.

* The oil will be refined in the United States and sold for some $50 billions.

CHAPTER 3

Future Energy Sources

3.1. Some Promising Energy Sources of the Future

The ideal energy source would be one having abundant raw material resources, operating without pollution, and at a cost lower than that of fossil fuels. How abundant a supply is needed can be estimated from the projected power uses of man. Assuming that the energy consumption will level off (the necessity for ceasing to strive for continuous growth will be discussed later) at a worldwide uniform rate of 50 kilowatts per person (five times the rate of the present U.S. use), and assuming that the earth's population will not go above 10 billion people, the total need will be 5×10^{14} watts. This is about one hundred times that of the present rate of use.

This is a tremendous demand and its bland denial would mean that there would always be rich and poor nations. What sources might we research and develop to produce this energy? One way would be the utilization of atomic energy in one fashion or another; another would be the possibility of capturing and converting to usual forms a portion of the energy radiated toward us by the sun. Some other nonpolluting energy sources are sometimes considered. They are geothermal energy (natural hot springs and steam), wind energy, tidal energy, and hydroelectricity. The energy available from these sources, however, is not abundant enough for man's projected needs; they could supply only about one percent of the 5×10^{14} watts. Of those which could provide an abundant supply, some are in a commercial stage; some are only a distant hope. These will now be discussed in more detail.

3.2. Atomic Fission, the First Step

Atomic reactors, producing energy by fission reactions, are familiar nowadays to most people. They are in the fully commercial stage and are being constructed all over the U.S.A. In this process, certain fissionable heavy elements (the raw material of the source) split up into smaller fragments, upon irradiation with neutrons, and release a large amount of radiant energy at the same time. Furthermore, very importantly, some of the fragments produced by the fission of the fuel are neutrons which can cause more fuel to react, producing more energy, and more neutrons, and so on. Thus, the process, once started, can continuously supply energy without the need for outside neutron radiation (Fig. 3.1). The energy generated is in the form of heat (the kinetic energy of the fission products), which is converted to electricity by one of the conventional methods, e.g., heat → gas

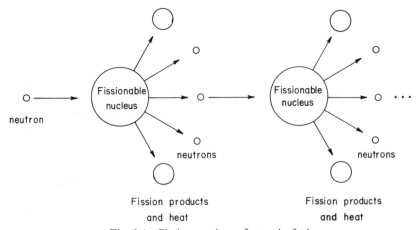

Fig. 3.1. Chain reaction of atomic fission.

expansion → force on piston → electricity from generator. The overall efficiency of the process is about 30%. The pollution from radioactive wastes, and the heat, can probably* be handled, as will be shown later, but there is a certain degree of atmospheric pollution from radioactive gases, the escape of which is probably impossible to prevent completely. The process is economically competitive with the old process at fossil fuel power stations. A problem is that the amount of the fuel supply (uranium) for these reactions is not abundant on the earth and would not last much beyond 2000. While it may seem, then, that this process hardly offers even a temporary solution to our problems, it is an important first step, since it leads to the breeder reactors described below.

3.3. Breeder Reactors, an Interim Solution?

While the supply of fissionable materials is limited, there are other raw materials which can be made fissionable by means of neutron irradiation. This fact forms the basis of the breeder reactor. The process is started with a supply of fissionable materials and is conducted in such a fashion that some of the neutron radiation produced is captured by materials which, not being fissionable previously, are made fissionable by the neutrons, and can, therefore, keep the energy production going (Fig. 3.2). As a matter of fact, there is a multiplication: while the fission reaction produces heat, and hence energy, it also produces more *fissionable* material than it uses in energy production (it can double its fuel supply in less than 10 years of operation). If this sounds, at first, much like *perpetuum mobile,* it is not.

An analogy: coal is presently being mined, processed, and transported with the expenditure of a certain amount of energy before it is converted to electricity at a power plant. The electrical

---

* Dealing with the waste products of the energy-producing fission process (which have a life of many thousands of years) looks increasingly, not decreasingly, difficult, as experience is gained. The life of the vessels in which they are contained is uncertain. Stress corrosion cracking is an unexpected difficulty.

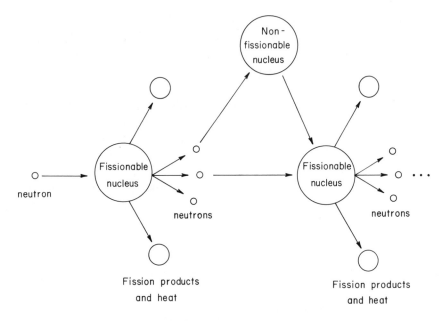

Fig. 3.2. The breeder. Some of the neutrons produced can be used to create new fuel.

energy so obtained is more than the amount used to furnish the coal, and parts of that electrical energy will be used to obtain more coal, which will make more electricity, and so on. The same with the breeder: parts of the energy obtained, in the form of nuclear radiation, are used to *process* further fuel, which will produce more energy and radiation, then fuel, and so on. But, in both cases, we are limited by the amount of energy stored for us by nature, in one case in the form of coal, in the other in the form of certain raw materials which can be made fissionable. The latter, however, have been estimated to last for at least 100 times as long as the fissionable materials (taking into account only the presently minable ores), promising us several hundred years of abundant energy (at a rate of 5×10^{14} watts).

Breeder reactors have been successfully operated, though only on a small scale, and with a price tag which is not yet competitive with fossil fuels. The basic problems are essentially solved, and the time for the development of the first generation of commercial breeders is expected to be less than 15 years. The cost of electricity generated by sufficiently large breeder reactors is predicted to be about half of that of fossil-fuel-based electricity (of course, in constant dollars).

The raw material situation for this means of energy production is worth discussion. Only 0.7% of all natural uranium is the uranium 235 isotope which is fissionable and which is at present almost exclusively the fuel of atomic reactors. About 99% is uranium-238, which can be made fissionable by neutron bombardment (turning into plutonium-239) and, therefore, can be used as a fuel in breeders. Thus, there is a tremendous increase in fuel utilization. But there is a further, even larger effect. Because a breeder reactor will utilize almost all of the natural uranium as fuel, against the less than one percent utilization of the "normal" atomic reactors, poorer uranium ores can be mined economically. Thus, if only all the uranium and thorium (which can also be made fissionable in a breeder) in the earth's surface crust could be utilized, our energy needs would be met for the foreseeable future (many millions of years?). The economics of recovery from deeper in the crust is of course likely to be extremely unfavorable.

3.4. Fusion, the Energy Source of the Stars

Energy is released during atomic fission—when a large nucleus is falling apart; nuclear energy can also be gained in the reverse type of process, when two small nuclei join to form a larger one. This is the type of process which supplies the energy of the sun. In stars, the energy comes from this process: the very large gravitational force of the enormous mass creates sufficient pressure to fuse two hydrogen nuclei to form helium. Several reaction schemes exist: the two most important ones are shown in Fig. 3.3. For such an atomic reaction to commence, the reaction mixture has to be heated to temperatures exceeding 50 million degrees. No material will stand these temperatures, so that, apart from finding some way of sparking off the reaction, it is necessary to find a very special way of holding the reacting mass. Because of

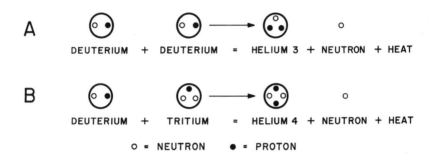

Fig. 3.3. Some fusion reactions.

the extreme temperature, the atoms become highly ionized and the reacting mass—a gaseous mass of ions at high temperatures— is called a plasma. The "container" considered to keep the plasma in place is a bottle-shaped magnetic field. Both the ignition and the magnetic containment have been achieved separately, but keeping the plasma stable for practical times is still a prospect for future research. There is no theoretical reason to suggest that the fusion reaction cannot be initiated and maintained in a controlled fashion, but the building of prototype laboratory reactors is probably several decades away. Correspondingly, it would be pointless to discuss economics at this time.

The deuterium–tritium reaction (Fig. 3.3B) could be realized most easily since it has the lower ignition temperature. In this case, there is a raw material limitation since tritium is rare in nature, but it could be produced by bombarding lithium with radiation in a "blanket" around the reactor. Considering the amount presently called minable, lithium as fuel would last only about twice as long as the materials for (nonbreeder) nuclear fission. There are, however, large amounts of low-grade lithium ores, the mining of which may become economical if fusion is achieved.

If the deuterium–deuterium reaction (Fig. 3.3A) can be realized (which is more difficult, requiring an ignition temperature about ten times as high as the previous reaction), the raw material situation will be drastically different. Deuterium can be obtained from seawater and its known recoverable amount would supply mankind with energy for more than a billion years, at a rate of 5×10^{14} watts. Thus, this type of power can be considered inexhaustible for man since the expected life of the solar system is also about 1 billion years.

There are further advantages of fusion over fission, in addition to the potentially inexhaustible supply of deuterium. It would be possible to obtain energy directly in an electrical form without the intervention of heat and heat engines with their low efficiencies. Such a scheme is shown in Fig. 3.4. The high-energy, charged particles leaving the reaction area could be slowed by an applied electrostatic field, and collected by appropriate electrodes producing high-voltage D.C. electricity. The electrons, collected at the first electrode, would travel through the load and back to the other electrode, where they would neutralize the positive plasma ions. The safety aspects of fusion are better. While a fission reactor can be *designed* to be safe, there is nevertheless increasing anxiety about the possibility of air pollution from breeder atomic reactors. A fusion reactor should be safer since the only possible escape of radioactivity is a leak of

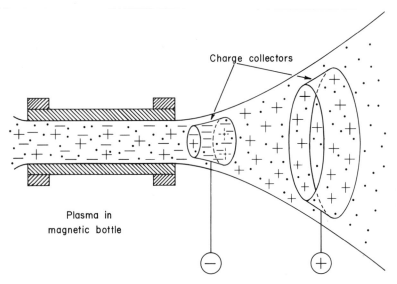

Fig. 3.4. Direct electricity generation from fusion.

radioactive tritium. Being a light gas, this would escape upward and be a lesser hazard than the radioactivity from the heavy atoms of the breeder reactor.

The sketch we have given of the possibilities from fusion energy may be too optimistic. Its use is still rather a utopian dream and not to be counted on to meet the 2000 A.D. energy crisis caused by the end of fossil fuels. Why not a NASA-like effort for fusion? Indeed, it might be worth it. But the problems of practical realization are so huge that it may be better to make such an effort in solar energy, where there are, also, no ecological objections. The greatest set of problems faced by fusion researchers lies in the immense difficulties of materials for use at the temperatures of a plasma: magnetic field might maintain it outside contact with materials, all of which would of course be instantly vaporized. But the extraction of electricity has to occur somewhere, and, there, contact has to be made. And what of radiation from gases at millions of degrees? An energy conversion method which depends upon maintaining a bit of artificial sun and drawing a current from it—that is utopian indeed.

3.5. Direct Capture of the Sun's Radiated Energy

The sun radiates 1.7×10^{17} watts onto the earth continuously. If we could convert this with an efficiency as low as 10%, only a small portion (3%)* of total radiation would have to be

---

* Capture of 3% of the sun's radiation would indeed be an upper limit. Suppose we can stretch the efficiency of collection on a given instrument to 20% (it is nearer 10% at present). Suppose we use not more than 1% of our land area for energy collection, then we should have 1×10^{14} watts, less than 1%. To get the 3%, and hence 5×10^{14} watts, captured, we should have to look at a second-generation solar collecting system, one we might contemplate having in the mid-21st century, perhaps. Here, one might contemplate extensive solar collectors on the sea—and the orbiting collectors which used to be science fiction and are now the subject of research financed by Arthur D. Little, Inc., the well-known American research firm. But the *need* to capture 3% assumes a superaffluent all-equal world population, which is certainly more than 100 years away.

captured to provide the 5×10^{14} watts projected as needed by man. Considering that this would be a truly inexhaustible source (and absolutely clean), the incentives for development are not small.

Why has not solar energy been made practical on a large scale at this time? Will not our descendants look back in dismay on our tardiness in the development of this resource? It has often been developed in tiny applications (hot water systems for individual houses), but a development in the direction of solar farms to produce electricity for a whole country has never been attempted. There are several reasons:

(i) The prospects from solar energy depend, firstly, upon the virtually certain availability of a sufficiently large number of kilowatt-hours per unit of territory. When this is worked out, it transpires that in the northern hemisphere, on land, the prospects are not very encouraging, for obvious climatic reasons. A low yield per unit area can be compensated by use of a bigger area of land, but the North European and North American areas are just those where land is never cheap and often exceedingly expensive.

In the countries which are rich in solar energy and have plenty of very cheap land—Australia, North Africa, bits of India—there is tragically a lack of that combination of scientific "big thinking" and enthusiastic far-seeing government to seize the opportunity which these countries do have by funding and organizing sufficient research to begin exploitation of their energy riches.

(ii) The economics of the process have been made to look very poor, and sometimes absolutely impossible, by quotations in the region of $100,000 per kilowatt for the basic plant.* This estimate has been arrived at by making the limiting assumption that only the photovoltaic method of conversion is going to be used. In this method, the light falls on a semiconductor which is arranged in two different forms to make up what is called a *p–n* junction. The photons from the sun enter a layer consisting of one type of semiconductor and, as a result of the interaction of

---

* The corresponding figure for fission reactors is about $500.

the photons with the semiconductor structure, a potential difference is formed at the junction between the p type and the n type. It is this potential difference—and the current to which it can give rise—which is the origin of the photovoltaic method of obtaining solar energy. Why was it thought to be so expensive? The reason lies in the necessary purity of the semiconductors. Unless they are superpure, the electrons activated by incident photons lose energy in collision with impurity atoms and become ineffective, i.e., the efficiency for the collection of the light and its conversion to electricity goes down. But—at least at present—the very pure materials needed are expensive. Further, while at present the gain which would be introduced by mass production would be several hundred percent, we need an improvement of about 1000 times. Very recent research by the Tyco Company suggests that a breakthrough in cost reduction has been achieved, although it still has to be confirmed by commercialization.

However, there are several other methods of transducing solar energy to electricity. In one, mirrors are arranged so that the sun is focused onto pipes containing liquids, e.g., sodium. If the system is covered with glass and vacuum introduced, cooling due to convection is enormously reduced and the liquid can be made to rise in temperature to several hundred degrees Celsius. Then, it can be made to flow through some conventional heat exchanger, heat water, say, and electricity can be made by the classical cycle of boiler and generator. The prospects of the cost per kilowatt for such an operation are in the region of those for atomic reactors.

This method may be made practical by concepts which depend upon the variation of the absorption (and emission) coefficient of a substance with wavelength. Thus, the sun's radiation is incident upon the earth at wavelengths which have their median value at less than one micron. The radiation which is reflected (after having lost energy due to absorption) has a wavelength of more than one micron. The net temperature which a substance attains upon being exposed to the sun depends upon the emission of some of the energy. In an idealization, if a substance had the characteristic that its emission (and absorption) coefficients depended on wavelength in such a fashion that they would be unity for wavelengths of less than one micron (i.e., absorb all the incident solar radiation) and be zero for wavelengths above one micron (i.e., emit none), then, if the substance were in a vacuum, it would continue to increase in temperature until the melting point. Partial achievement of this concept has recently been reached. The further development difficulties are again material ones. How long will the material, with the characteristics mentioned, exist without substantial decay?

An improvement in this method is to seek substances which possess high absorption coefficients and low emission coefficients (usually, they are nearly equal). If such can be found, a temperature of 300–400°C can be reached in the central metal.

Another possibility is to use thermocouples to transduce the heat to electricity. The efficiency of these devices is poor at lower temperatures—say, 200–300°C—but if the temperature can be raised to 400–500°C, the efficiency can be increased to more than 10% and perhaps to 20%. The latter efficiency would be somewhat greater than could be obtained by the use of photovoltaic cells and about equal to that for the chemical cycle via steam. But in this latter case, there is no need for a flowing liquid or for generators. And, above all, there are no expensive materials needed.

So government people have been put off funding research on solar energy by concentrating attention on the former negative cost feature of solar cells. But the possibilities of other methods are quite real, and the development of this resource, particularly for countries in the southern hemisphere, seems one of the greatest attractions in applied science at any time.

3.6. Some Comparisons and Conclusions

How do these possible future power sources measure up against the earlier stated desired characteristics of abundant raw material, no pollution, and low cost? Table 3.1 summarizes the raw material resources. It is clear that, even with conservative estimates, both the breeder and the fusion reactor would enable us to continue civilization for a long time. The sun's radiation is

TABLE 3.1.   **Estimated Energy Resources**

Source	Estimated supply 10^{12} watt-years	Years of supply	
		at present rate $(5 \times 10^{12}$ watts)	at a rate of 5×10^{14} watts
Fossil fuels	1300	260	2.6
Nuclear fission	3000	600	6
Breeder reactors (using presently minable fission materials)	300,000	60,000	600
Total fissionable materials in the earth's crust	1.6×10^{12}	320×10^9	3.2×10^9
Lithium for fusion (presently minable)	7000	1400	14
Lithium for fusion (total in the earth's crust)	2.2×10^{12}	440×10^9	4.4×10^9
Deuterium for fusion in the sea	6.6×10^{11}	130×10^9	1.3×10^9
Sun's radiation	1.7×10^{17} watts (continuous)	No limit	No limit
Tidal energy	10^{12} watts (continuous)	Not sufficient	Not sufficient
Geothermal energy	6×10^{10} watts (continuous)	Not sufficient[a]	Not sufficient

[a] The estimate refers to surface sources and not to tapping of the heat of the magma 40 miles below the earth's surface.

a sufficiently large and attractively inexhaustible supply of energy. These conclusions were based on an assumed leveling off the total energy consumption at a rate of 5×10^{14} watts, a figure about a hundred times as large as the present rate. The reality of this assumption will be discussed in the next section.

Environmental pollution is a criterion, however, which would favor one source over another. Although no noxious fumes, no smog, and no carbon dioxide would be produced from atomic reactors, the breeder brings up the problem of handling of radioactive wastes. Though they are very dangerous, they have a small total volume and, therefore, feasible disposal methods should not be too hard to find. Deposition in deep mines is considered and, as a potential final solution, ejection into space and to a solar orbit may not be unrealistic one hundred years from now. Another area is nuclear safety; the processes taking place in the reaction are the same type as those of the atomic bomb. Could a reactor "run away," causing a nuclear explosion? Reactors are designed with extensive safeguard features, and, as a result of this, there have been no major accidents in the quarter-century history of the process. When accidents did occur, they were confined to the reactor area and caused no harm to the population. The most serious problem is the escape of radioactive xenon and krypton, which would greatly increase leukemic cancer in the population. The fusion process is free from the explosion problems, but generates radioactive tritium which could be harmful in large quantities.

There is another problem—the ejection of waste heat. This occurs with most energy-producing techniques where heat is converted into electricity, including the fossil fuel plants, a process having an inherently low efficiency. The reason why it will be more of a problem in the future is the great size of the new power plants, concentrating large amounts of heat into small areas. The large size is dictated partly by the inherent relation between efficiency and size of nuclear plants. The heat, however, is not an insurmountable problem, though it places severe restrictions on the location of the new power plants. The only acceptable location seems to be near or on large bodies of water (oceans, large lakes), where abundant cooling water will be available. To preserve shore space for other uses, the power plants will probably be forced onto the water, possibly onto floating platforms quite a few miles offshore (to have sufficient available cold water for pumping up and cooling). The heat produced will be dissipated into large volumes of water, causing only small local temperature changes.

With regard to economic and technical realization, the breeder is ahead of other schemes. It will reach commercial size in the 1980's. Were the proper funding to be given, it is possible that we could build breeder reactors fast enough to supply more than half our energy needs by the end of the century. Fusion has a much less certain time scale for development; possibly it could be 50 to 100 years.

One of the important questions of the time is the relative desirability of developing solar energy compared with breeder atomic reactors. The main point in favor of breeders is that they are likely to be actually usable in the 1980's (though only a few reactors will have been built by then), whereas unless a real Manhattan Project effort were made with solar energy, it probably could not be available for large-scale use until 1990 or 2000. However, neglecting this factor of happenstance, that the development of atomic energy got under way on a big scale more or less by misunderstanding of the cost aspects of solar energy, all the other factors seem to favor the latter. There still hangs over us the nightmare of pollutional hazards from the breeder. It would be disastrous indeed to go down the next few decades building the breeder too quickly—in the fear of fossil fuel exhaustion or pollution—only to find that the genetic damage to the population was too high a price to pay.

Two very considerable advantages would accrue from the large-scale development of solar energy. On the one hand, absolutely no pollutional hazard exists. But, on the other, there is no great new input of heat to the earth. The sun's rays arrive anyway, and if we make electricity by transducing the heat, there will be no thermal consequence.

One concern with solar energy is that it is much more abundant on land areas and oceans far from the present centers of production. This problem is not unsolvable, however, because the electricity from solar energy can be converted to hydrogen and then shipped, just as natural gas is today. On arrival in a given country, it can be distributed in gaseous form by pipeline.

The area of (say) Australia needed to supply 300 million Americans with 10 kilowatts per head is a patch about 200 miles square.

In summary: So long as the population is limited to the figure assumed, we *could have* eventually a plentiful supply of energy for everybody at a highly affluent level. There are two possible sources to be considered for development: breeder reactors and solar energy. The breeder is much more developed than are converters of solar energy, but does have terrifying pollutional hazards. The development of solar energy has been held up because of a now outdated concept of the cost picture.

An immediate considerable acceleration of research on both these sources (as well as the further-off fusion process) seems necessary. By a date as near as 1980, several substantial increases in the price of oil and natural gas will likely have occurred, and after 1985 we will enter into a dangerous period (until the new sources are practical) in which the energy supply of the United States will derive exclusively from the Arab states (and the Soviet Union).

A retreat to burning coal once more as an energy source is impractical. Burning it without removal of sulfur would result in unacceptable pollution, and removal of the sulfur would seem too expensive at present. And there is the danger of climatic changes discussed in Chapter 2, which would arise if we took seriously the retreat to coal.

3.7. Can We Use Unlimited Amounts of Energy? The Heat Pollution Problem

Earlier we have dismissed the heat problem created by large power plants by locating them near or on large bodies of water. While that is correct for each individual plant, one also has to consider the sum of the effects upon the overall heat balance of the earth. In this case, however, not only the waste heat of the power plants, but all energy used by mankind should be taken into consideration. In whatever way we are consuming energy (transportation, chemical processing, mechanical work, heating, lighting, etc.), practically all of it will be degraded to heat during the process or not too long afterwards. In the overall heat balance calculation, therefore, the total energy consumed by man should be used. The waste heat of the power plants comes in through their efficiency; if they operate at about 30%, the total energy used by man will have to be multiplied by three to obtain the total heat added to the earth.

This is how the 5×10^{14} watts leveling-off value, used throughout this chapter, came about. In the heat balance, one has to compare the total amount of energy received from the sun to that generated by man. It has been estimated that the earth's temperature will rise by $0.1°C$ if man adds two-tenths of one percent of the heat received from the sun. As was shown in Chapter 2, a few tenths of a degree change in the earth's temperature can be tolerated, but when we start to talk about degrees, it may mean disaster. The 5×10^{14} watts, produced with about 30% efficiency in the power plants, means 1.5×10^{15} watts added to the earth, which is only 1.2% of the 1.2×10^{17} watts received from the sun (the total energy of the sun's radiation reaching the earth is 1.7×10^{17} watts, but 30% of it is reflected back as light). The 5×10^{14} watts level, therefore, seems a sensible upper limit which would result in less than one degree of expected temperature increase.

It should be noted that the efficiency of power generation may become critical at this point. If instead of the $30-40\%$ efficiency of the heat engine we could operate at the $60-80\%$ of direct electricity generation, our maximum allowed energy quota would double. If we developed *solar* conversion, no extra heat would be added to the earth's load. The earth reflects back a certain amount, on the average 30%, of the sun's radiation as light. Based on this figure (which will change from place to place), if the power plant is engineered to reflect back 30% of the energy as light and absorb the remainder, no extra heat has been added to the earth's balance. The absorbed 70% will all be degraded to heat, some of it at the power plant, the rest at the point of use, but this is the same 70% which would have been added to the earth's heat from the sun's radiation in any case. This again could increase our quota, though not indefinitely.

The conclusion is inevitable that, at some level, the total energy used by mankind will have to be limited. The notion that the population growth on the earth will have to level off in the near future is fairly generally accepted; it should not be much more difficult to live with the concept of limited power. What about the maximum level? How would it affect our standard of living? If power consumption can be directly related to standard of living, the 5×10^{14} watts limit considered here will allow a population of ten billion to live at a uniform standard of living five times as high as that of the United States at present. This is not a bleak future; it is a materialist's paradise.

3.8. Some Common Characteristics of the Future Energy Sources and Their Consequences

Whether present social structures will allow man to realize these possibilities, and what social effects such an abundant, energy-rich life for all would have, are outside the scope of this book. We are, however, concerned with the technological consequences of the absolutely necessary move from the burning of fuel to atomic or solar sources. These will follow from some common characteristics of the possible new power sources and, therefore, are valid whichever will become the final choice.

The power centers of the future will be huge and far away from a large portion of the population. The atomic installations will have to be large to achieve a sufficiently low cost per unit energy and, because of the large amounts of waste heat, and possibly for safety's sake, they will be located on offshore islands. Solar power stations will be concentrated at geographical locations providing the most uniform sunshine. *All the power generated at these remote stations will be in the form of electricity,* which will have to be transported (sometimes many thousands of miles), distributed, utilized efficiently, and mobilized (for transportation). The most important single technical consequence will be that the technology of the future is electrically, and therefore often electrochemically, oriented.

Electricity, as such, can easily be transported in wires, but this starts to become expensive when thousands of miles are involved. It can be shown that transformation of electricity into chemical energy (production of a synthetic fuel) and transportation of this fuel, with reconversion into electricity at the point of use, is more economical. Electricity, as such, cannot be conveniently stored. But a reversible transformation between electricity and chemical energy is practical. Such storage of electricity (e.g., in hydrogen as a result of the electrolysis of brackish water) is vital for smoothing out the load of the power plants over the peaks and valleys of the use pattern, and it is also important in transportation. It is not convenient, in most cases, to run electrical energy in wires to mobile units; electricity will have to be converted to chemicals and later reconverted in the vehicle. But every time we are talking about converting electricity into chemical energy or using chemical sources to give electricity, we are talking about *electrochemistry* (see Chapters 4 and 5).

But that is not the entire extent of electrochemistry. Most chemical reactions, if not all, can be run electrochemically if desired, and some of them already are. For some reactions, the economics *of the present* are such that, since coal is still cheaper than electricity, it is advantageous to proceed by heat. But when fossil fuel exhaustion and/or pollution has caused an all-electric economy to be set up, it will be more economical and pollutionally acceptable to use electricity as such, rather than producing heat over a resistor to drive something. The chemical technology, therefore, which is involved in all basic needs of man, such as food (preservation, fertilizers), clothing (synthetic fibers), shelter (building materials), health (drugs), etc., will become, in a society in which the medium of energy is entirely electricity, a largely *electrochemical technology*. Moreover, the extraction of metals from ores will gradually become a thing of the past. All metals will be recycled, and separated out from others with which they have been alloyed, by electrochemical means. The clean-up of factory liquids will be achieved electrochemically. Factory water will be recirculated, and electrodialytically regenerated. Sewage will be electrolytically processed. Cheap and abundant (electrochemically produced) hydrogen will be used on a large scale for chemical purposes. Finally, for the long term, the only pollutant which can be allowed in an ecologically viable world is heat (and not too much of that). Atoms (or solar energy) will generate electricity; hydrogen will transport it to the local site, and fuel cells will regenerate it; electricity will operate technology without pollution by means of electrochemistry; and the final product will only be heat. All else must be recycled, and it seems likely that much of that will be electrochemically contrived as well.

CHAPTER 4

Electrochemical Sources of Power: Batteries

4.1. The Need for Electrochemical Power Sources

In the all-electric economy projected for the future in the preceding chapter, the very extensive use of electrochemical power sources will become a necessity. There are a large number of applications, ranging in size from the powering of a pacemaker for a heart to the powering of a truck, where large and remote primary power sources cannot be used, nor can electricity be supplied by wires. Electrochemical power sources can provide the electricity for these applications. Emergency, standby power sources, to be used in case of power failure in the primary system, will also have to supply power in the form of electricity. Another area is the storage of "off-peak" electricity. For the most efficient use of a power generating plant, a uniform load distribution over the 24 hours of the day would be needed. This is especially true for breeder atomic reactors, where "off-peak" electricity can have a negative cost. The breeder produces more fuel than it uses, but only when it operates. For the continuous operation of power plants near full capacity, an energy storage system is required to smooth out the hills and valleys of the use pattern, an important use for electrochemical storers.

47

The ecological uniqueness of these power sources is important. A battery does not emit byproducts. The emission from a fuel cell operating on hydrogen and oxygen is only water. Their efficiency is in the 60–80% range, compared with the 20–30% of conventional power sources, so that waste heat production would be reduced. These characteristics, coupled with their compatibility with an all-electric economy, will make their use widespread.

Most technologists outside the electrochemical field are not aware of the advances in electrochemical power sources made since the 1960's, and that they are comparable to (in terms of power per unit weight) or better than (in terms of energy per unit weight) conventional ones.

4.2. There Are Two Types of Electrochemical Power Sources

Although we have spoken of fuel cells and batteries, it is necessary for us to distinguish clearly between these two types of electrochemical power sources, because they have different aims.

(i)   The purpose of the battery is to take electricity, *produced by another source* (e.g., a solar collector), and store it for later use. It could be called an "electrochemical energy storer."

(ii) The purpose of the fuel cell is to take a chemical fuel (e.g., hydrazine) and *create* electricity in the fuel cell through the electrochemical reactions of the fuel with oxygen from the air. This is a true *electrochemical power plant*.

Thus, if we had limitless amounts of fuel, and did not have the prospect of limitless atomic energy and electricity from fusion and solar power, we should concentrate on developing fuel cells.

However, abundant electricity from breeders and solar energy is probable, and so the bulk of our research in electricity storage should be to develop high-energy and high-power density batteries for use, e.g., in automotive transport, and fuel cell systems capable of being used as electricity storers (*cf.* the hydrogen economy, Chapter 6).

Batteries will be discussed in this chapter, and fuel cells in the following chapter.

4.3. The Principle of Batteries

In an electrochemical electricity storer, electrical energy is converted into chemical energy which can be, at will, reconverted into electricity. The reason why such energy conversion is possible is that in some chemical reactions (the so-called oxidation–reduction reactions) an electric charge transfer from a molecule or atom to another takes place. For example, mixing zinc powder into a solution of copper sulfate will result in the dissolution of the zinc and the precipitation of copper powder:

$$Zn + CuSO_4 \rightarrow ZnSO_4 + Cu$$

or, considering that the salts in the solution are ionized and the sulfate does not take part in the reaction,

$$Zn + Cu^{++} \rightarrow Zn^{++} + Cu$$

This latter reaction shows that during the process positive charge is being transferred from the copper to the zinc, or negatively charged electrons are transferred from the zinc to the copper. The zinc is being "oxidized" (its positive charge is increased), and the copper is being "reduced" (its positive charge is decreased).*

Useful electrical work can be obtained during the reaction if the process of the zinc giving up its electrons and that of the copper picking up the electrons can be spatially separated and the electrons *en route* from the zinc to the copper led through an electrical machine, e.g., the armature of a motor. This can be achieved in the system shown in Fig. 4.1. Copper metal immersed

---

* The term "oxidation is of historical origin; a material combining with oxygen increases its positive charge. "Oxidation" without actual reaction with oxygen is possible in the broader sense defined above.

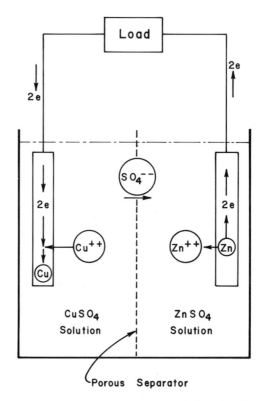

Fig. 4.1. Energy-producing process of the copper–zinc cell.

in copper sulfate solution and zinc metal immersed in zinc sulfate solution are separated by an ion exchange membrane which allow the passage only of anions (in this case sulfate), but not that of the cations (metal ions), and prevents the mixing of the solutions. What happens is that the dissolving zinc will leave two of its electrons in the metal. These electrons will then travel through the load into the copper electrode and combine with a copper ion to form copper metal:

$$Zn \rightarrow Zn^{++} + 2e$$
$$Cu^{++} + 2e \rightarrow Cu$$
$$\overline{Zn + Cu^{++} \rightarrow Zn^{++} + Cu}$$

The sum of the two electrode reactions is the same as the chemical

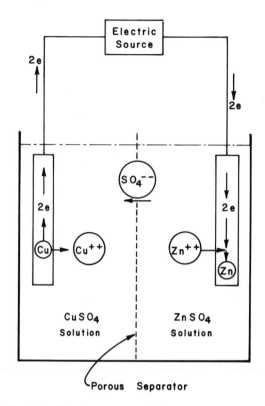

Fig. 4.2. Charging process of the copper–zinc cell.

reaction written above, but now the reduction and the oxidation are taking place at separate locations and the electrons traveling from one site to the other can be used for useful work, Such a "power source" will continue to supply electricity until all the zinc is dissolved or all the copper sulfate is used up, whichever happens first. The load shown in Fig. 4.1 could be any electric device, a motor, a heater, a light bulb, etc.

This is then the electrochemical power source converting the energy of a spontaneous chemical reaction into electricity. How can this now be used for energy storage? If, after the battery was "discharged" (the available electricity used), in place of the load one puts an electric generator and forces electrons backwards (Fig. 4.2), the following reactions will take place:

$$\begin{aligned}
Zn^{++} + 2e &\rightarrow Zn \\
Cu &\rightarrow Cu^{++} + 2e \\
\hline
Zn^{++} + Cu &\rightarrow Zn + Cu^{++}
\end{aligned}$$

which is the reverse of the power-producing reaction. In this way the state of the battery can be restored (the battery is "charged"), and it will again be able to produce electricity upon connection of a load. The cycle can be repeated indefinitely. The discharge of the battery does not have to take place immediately after charging; the battery can be kept in the "charged state" (zinc in the solid form and copper in solution in ionic form in our example) and the electrical energy stored in it for future use at any time* and any place. This then is an electrochemical energy storer.

The copper–zinc reaction was only taken as a simple example. A more practical cell is, for example, the zinc–silver cell. The overall reaction is

$$Zn + AgO \underset{\text{charge}}{\overset{\text{discharge}}{\rightleftarrows}} ZnO + Ag$$

Zinc is oxidized (literally in this case) during the discharge by the silver oxide, and the reverse happens on charging. The reactions taking place on the separate electrodes are complex in this case and cannot be given as simply as for the zinc–copper cell, even though the overall reaction is represented simply. In the battery, the electrode materials are present as a porous mass (to provide high surface area) pressed onto a supporting metal grid. These electrodes (plates) are then separated by cellophane membranes and immersed in the solution (concentrated potassium hydroxide). A schematic drawing is shown in Fig. 4.3.

The efficiency of the storers depends on the specific system in question and on the type of use; it is rather high (60–90% range). There are some charge losses due to side reactions during

---

* This is an ideal statement. In fact, there are always side reactions, corrosion. Batteries left unused for years lose energy and gradually discharge themselves.

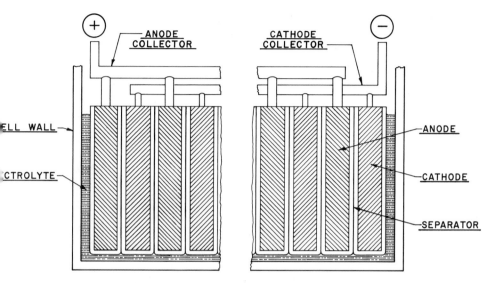

Fig. 4.3. Schematic of a silver–zinc battery.

charging and discharging, and on standing. Further, the electricity is returned at a lower voltage level during the discharge than the voltage which was used during the charging, causing additional losses. These so-called overpotential (or polarization) losses will be treated in more detail in Appendixes II and III. A typical charge–discharge curve is illustrated in Fig. 4.4, showing the change of battery voltage during operation.

4.4. Requirements for Electric Transportation

While numerous uses for electrochemical power sources can be foreseen in an all-electric economy, there are reasons why transportation should be emphasized. About one-fourth of our energy is spent on transportation, and this ratio will remain unchanged for the next few decades. Further, this segment of our energy consumption is the main cause of pollution of the atmosphere, especially in urban areas. Because of the pollution situation, electrification of all our transportation may be necessary before the advent of an all-electric economy.

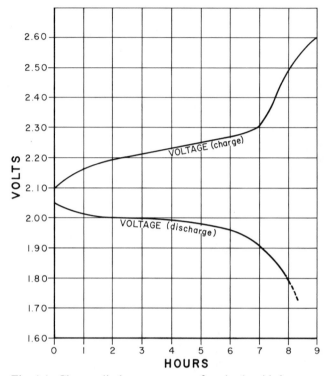

Fig. 4.4. Charge–discharge curves of a lead–acid battery.

Transportation will emerge as an ecologically important and large-volume application for electrochemical power sources. Another reason for considering transportation requirements is that they are stringent. The requirements of a power source used to provide electricity for a household are less severe as far as weight, volume, ability to withstand extremes of temperature, mechanical handling, etc., are concerned. Transportation is the acid test of a small power source.

Requirements for a family car will be discussed, because they are the most demanding among the ground transportation means. The requirements arise from the limitations of the percentage of weight and volume of a car which can be allocated to the power source, and from the minimum energy and power needed during use. Energy and power are taken separately because they relate to different needs, and place different requirements on the power source. The energy requirement is determined by the length of travel desired between "fill-ups." If this is taken as a 200 mile range, the energy density requirement is 120 watt-hours/pound (WH/lb). This in itself does not fully characterize the power source since it makes a large difference how fast the energy has to be supplied (120 WH can be used, e.g., at the rate of 120 watts for 1 hour or at the rate of 1 watt for 120 hours); the needed rate is determined by the minimum power requirement. For a family car, this is estimated to be 100 W/lb. This results not from the desired speed (for a 70 mph cruising speed only 40 W/lb is needed), but from acceleration requirements and the necessity to maintain speed uphill. The acceleration requirement is to reach full speed in 20 seconds. While this may seem excessive, it is a necessity under today's traffic conditions (for example, entering an expressway).

The requirements can be summarized as 120 WH/lb and 100 W/lb. There are additional requirements, such as price (not much greater than those of the dirty old sources) and safety (e.g., a cold battery would be preferred to one operating at several hundred degrees).

4.5. Characteristics of Present Batteries

The energy and power density achieved by some *present* batteries are given in Table 4.1. None satisfies all the requirements of a family car at present, though the characteristics of some are not too far from the goal.

The lead–acid battery (more than a century old), familiar to everyone owning a car, has the poorest performance of all in energy density* and is no contender as the principal electrochemical power source of vehicles. Several of the more recent developments show more promise. As far as *power* density goes, the requirements can be met and surpassed; energy density, rechargeability, and price are the areas that need to be

---

* It is, however, relatively cheap.

Table 4.1. Characteristics of Present Batteries

Type	Energy density WH/lb	Power density W/lb
Lead–acid	15	80
Nickel–cadmium	25	250
Silver–zinc	50	150
Nickel–zinc	25	200
Manganese dioxide–zinc	25	100
Air–zinc	80	35

improved. The energy density of batteries so far developed is only about 20% of the theoretically calculated maximum. Some of this loss is unavoidable since the efficiency is not 100% and some packaging and structural material is needed to hold the active electrode materials together, but there is room for improvement by research. The rechargeability of some systems is less than satisfactory. One would like to be able to take a battery through several thousand charging–discharging cycles* (various troubles arise at a greater number, e.g., the plates start to disintegrate), but some, using, e.g., zinc electrodes, are good only for a few hundred. This makes their use too expensive. The use of expensive materials, like silver, is prohibitive not only because the amortization costs are too high, but because there is not enough silver available (unless it could be extracted from sea water). Silver–zinc cells might have satisfactory characteristics for some exotic types of cars (perhaps for fire engines and ambulances where high-speed dashes over short distances and instant power (no warm-up period) are important.

The air–metal combination is a newly developed system of value. It is half battery, half fuel cell, since one of the active materials (oxygen) is supplied continuously from the outside, from the surrounding air. The reason for its good energy density rests partly on this fact, since it has to "carry with itself" only one half of its active material. While the use of air gives an advantage with respect to the energy density, it is also the

---

* For cars, with a once-per-day recharging process, at least 1000 charges are necessary for *ca.* 3 years of life.

Table 4.2. Expected Characteristics of Emerging [a] Batteries

Type	Energy density, WH/lb	Power density, W/lb
Lithium–chlorine	180	180
Lithium–sulfur	190	130
Sodium–sulfur	150	160
Organic	110	20

[a] Emerging means batteries which have emerged from research and are in the development process but have not yet been made commercial.

reason for the low power density because of the poor overpotential characteristics of the oxygen electrode (this will be discussed further in connection with fuel cells; see also Appendix II).

Thus, while none of the present batteries is fully satisfactory for the electric family car, some of them are close enough to give support to the expectation that, by further research and development,* they can be brought up to the required performance level.

4.6. Some Promising Batteries Presently in Development

Because of the necessity of abandoning fossil fuels as the basis of our energy resource, some research has been carried out to develop new systems. The characteristics of some of these are shown in Table 4.2. The common feature is that they do not use water as a solvent. This is because, to improve the energy density of the storers, highly reactive electrode materials are used which can provide increased amounts of charge per pound at high voltages. These materials (e.g., sodium, lithium) cannot be used in aqueous systems since they would react instantaneously with the water.

* The research is primarily in electrochemical engineering. However, there are quite a number of problems which involve studies in fundamental electrochemistry. For example, we know little about how crystals grow under the electric field strength in the double layer.

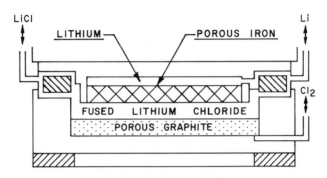

Fig. 4.5. A suggested design for the lithium–chlorine battery.

The lithium–chlorine cell (Fig. 4.5) has a molten salt electrolyte, one molten metal electrode, and one gas electrode. The reaction is simple:

$$2Li + Cl_2 \underset{\text{charge}}{\overset{\text{discharge}}{\rightleftharpoons}} 2LiCl$$

The construction is not so simple because of the highly reactive nature of the molten materials and the chlorine gas. The reacting materials have to be stored outside the cell proper, increasing the complexity. The expected performance characteristics, however, are so good that serious efforts to overcome the construction difficulties and corrosion problems are worthwhile (though at present little pursued).

The sodium–sulfur battery (Fig. 4.6) uses a solid ceramic as "electrolyte" (which is conducting at the temperature used) and molten sodium and sulfur as electrode materials. It would be self-contained, like a "normal" battery.

These high-temperature batteries (operating at 300–700°C) have inherent drawbacks. They would require additional heaters for start-up and might need a short warm-up period. The construction is difficult because of heat insulation and safety considerations. It is the high temperature which also gives them their excellent performance characteristics because of the increased rate of reaction which it brings (this implies a smaller reaction loss due to overpotential, *cf.* Appendix II) and decreased electrical resistivity of the electrolytes.

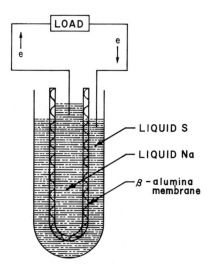

Fig. 4.6. Schematic of a sodium–sulfur battery.

Another approach is to use organic liquids as electrolytes in the battery. Here again, reactive materials not compatible with water could be used, *and* at low temperatures. Most salts are not highly soluble in organic liquids, and this produces high-resistance solutions with (consequently) low power efficiency.

These examples suffice to show that new systems, capable of meeting the stringent requirements of a family car, could be evolved, given sufficient research and development effort. The fact that the very short and limited research efforts of the last few years could produce the promising results of Table 4.2 suggests that the realization of the characteristics needed in a commercial and practical form is possible.

4.7. Some Obstacles to Be Overcome in Developing the Needed Battery

A number of research and development tasks have to be done to achieve the promise of batteries shown in the last section; most of these are engineering development problems, some are fundamental.

(i) Materials

Many of the newer designs of batteries, in particular those which work at high temperatures, are held back in development by the absence of research in materials science. The requirements are high considering the reactivity of the electrode materials, the elevated temperatures, and the safety precautions needed for the family car. This is an area where, with a little conscious effort, some "spin-off" from space research is possible. However, these matters are under-researched and provide the principal difficulty to progress.

(ii) The Weight Problem

This is partly connected to the first problem. The structural materials are part of the battery package and, therefore, in addition to their corrosion resistance, they should be lightweight to help achieve the desired energy and power densities. There is another side to the problem, the weight added by the reactive materials, if they are not utilized to the fullest possible extent. Thus, in present batteries, the individual "plates" containing the reactants cannot be fully discharged before being charged. While further improvements are possible, this is inherent in the plate-type design, where the starting and end materials are intermingled on the same electrode. Different design concepts (e.g., the external reactant storage of the lithium–chloride battery) are needed for improvement.

(iii) Ability to Recharge Repeatedly

One of the necessities of a battery is that one should be able to recharge it many (*ca.* 1000) times. With some batteries, side processes take place which make the material in contact with the solution less and less available with each cycle of charge and discharge. For example, the crystal particles of the chemical on the plates of the battery aggregate together to form big particles in which more of the active material is on the inside and no longer in contact with the solution. This also ties in with the weight problem since every particle present which does not react with the solution in the production of current is just a weight burden.

4.8. Nonrechargeable Batteries

Everyone is familiar with today's throw away "dry cells" used in flashlights, transistor radios, and many other gadgets. They are often called batteries, but they are really a hybrid between fuel cells and batteries. They are like fuel cells since they convert the chemical energy of a fuel to electricity. They are not storers of electrical energy since they are not rechargeable by it. But fuel is not fed to the dry cell continuously, contrary to the situation in a regular fuel cell; fuel is placed inside during the manufacturing and the cell operates until it is all used up. These cells are a self-contained closed system; and in this respect they are like batteries.

From an ecological standpoint, they are wasteful, using the chemical fuel only once, and they will be replaced by rechargeable batteries. Already there is a tendency by manufacturers to make them partially rechargeable (some can be reused a dozen times). The dry cell will not disappear completely, however, since it can be made very compact and light, important features for some applications.

4.9. Electricity Storage with Fuel Cells

Fuel cells can be used for storage purposes. In case of a power plant, which is required to produce electricity at a continuous, uniform rate to keep its efficiency high, at night, i.e., during the "off-peak" hours, the surplus electricity can be used to produce a "synthetic fuel" by electrolysis. The fuel can be stored and fed, during the peak demands of the next day, to fuel cells for reconversion into electricity. Many different schemes could be proposed; the one which follows was selected because it shows a blurring of differences between batteries and fuel cells.

Lithium chloride could be electrolyzed to produce lithium metal and chlorine gas to be stored separately until they are recombined in a fuel cell. This is a system very similar to the battery shown in Fig. 4.5. What then is the difference between batteries and fuel cells? *Electrochemically*, only that the reactants are held outside the electrode plates (often in tanks) for fuel cells, but in crystalline form on the electrode for batteries. There are differences in the practical objectives and, therefore, in the engineering to achieve them. If the objectives are the same, as in this case, electricity storage, these differences may get blurred, though they still exist. The electrolysis unit to produce the lithium and the chlorine for storage, and the fuel cell to reconvert them to electricity, would be two separate physical entities at the central power plant. The reason is to make each individual conversion step as efficient as possible. There are no serious weight and space limitations for these units, but energy losses should be minimized. On the other hand, for an electric car battery with the same electrochemical system, the same two electrodes in the battery will convert LiCl to lithium and chlorine and back to electricity. The gain in efficiency from the two separate units would not justify the extra weight and volume, which are at a premium in a mobile unit. This is an example of the common situation when advancements of two separate fields produce a breakdown of boundaries between them. (The difference—fuel cells *convert* the energy given out in electrochemical reactions to electricity; and batteries *store* electricity produced by some other device—is diminished if the fuel cell's fuel is produced by the fuel cell itself using electricity produced by other primary sources.)

CHAPTER 5

Electrochemical Sources of Power: Fuel Cells

5.1 The Principle of Fuel Cells

It is easy, if we forego sophistication, to explain how a fuel cell works. Let us look at Fig. 5.1. The fuel (let us exemplify using hydrogen) enters and is impelled into contact with one electrode in solution. It dissociates, and the resulting hydrogen atoms adsorb upon the electrode. They then give up their electrons to the metal (this is the fundamental electrochemical act of creating electricity), and these electrons then flow around the circuit, *through a load*, until they reach the other electrode. Here, they transfer to oxygen, which then, along with some protons from the solution, undergoes successive reactions, until water is formed. The overall fuel cell reaction involves oxidation of the fuel and reduction of oxygen occurring on separate electrodes. The stream of electrons is taken from the hydrogen reaction sites to the other electrode to react with oxygen through an outside load where this current is used to produce useful work (e.g., by passing through the armature of a motor).

The similarities between fuel cells and batteries have been stressed before; it is time now to emphasize the differences between them:

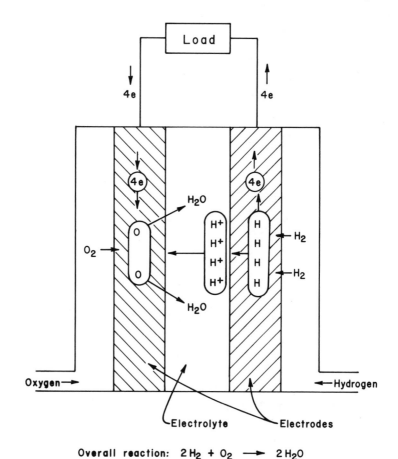

Fig. 5.1. The principles of the hydrogen–oxygen fuel cell.

(i)   The fuel cell is unidirectional, it produces electrical
      energy by the oxidation of a fuel (somewhat similarly
      to a conventional power plant). The reaction in a
      battery is reversible at will. It can proceed in either
      direction: it can produce electricity, or it can accept
      it for storage.

(ii) The fuel cell is supplied continuously, from the out-
      side, with a flow of fuel from a tank and with oxidizing
      material, e.g., air (again rather similarly to a conven-

tional power plant). The battery contains all the reactants within itself, it is a closed system with no material feed; only the electricity flows in or out.

(iii) As a consequence of (ii) the reacting materials in the fuel cell are usually gases or liquids, and seldom solids. On the other hand, those of the battery are usually solids, possibly liquids, and seldom gases.

Summarizing briefly: The electrochemical principles of batteries and fuel cells are the same; their intended use, and therefore, their engineering are quite different. The fuel cell works very much like a conventional power plant in that it accepts fuel and converts the chemical energy of its reaction with oxygen into electricity. The battery accepts electricity and stores it for future use.

5.2. Some Fuel Cell Systems

There are large numbers of chemical reactions which can be used as a basis for fuel cells, using a variety of fuels and oxidizing agent, and there are several engineering approaches to bring the process to realization. It would serve no purpose to enumerate or classify them; their description can be found in standard textbooks. Some examples will be given only to indicate the overall scope of the field.

Air (or oxygen) is usually used as the oxidizing agent. In principle, other materials could be used, and many have been suggested and tried. The advantages of air, however, are obvious: it is available at low cost, avoiding the necessity of carrying both the fuel and the oxidizing agent in separate tanks. It is, therefore, used in most systems suggested for widespread use.

The fuel systems are more varied. Hydrogen was already mentioned. The hydrogen–oxygen (air) system is a unique one for several reasons. It is the most developed (engineered) fuel cell, with many different designs being available. It is the system which has already had practical use, and publicity, through space exploration. It is also unique because it would be the heart of the proposed hydrogen economy discussed in Chapter 6. This system will, therefore, be treated separately in some detail later.

A large group of possible fuels comes from petroleum products, from derivatives of mineral oil and natural gas. These hydrocarbons could be fuels in the usual sense. A number of different systems have been developed. They have one common disadvantage vis-à-vis synthetic fuels (such as hydrogen): their rate of "burning" in the cell is slow (high polarization losses, cf. Appendix II). Another disadvantage is the fact that they still emit carbon dioxide, though to a lesser extent than in conventional power plants because of the higher efficiency of the fuel cells. The most classic fuel, carbon (coal) itself has been suggested and tried with limited success. (The name "fuel cell" comes, indeed, from the original suggestion of "burning" classical fuels electrochemically). The electrochemical reactions taking place in these devices are complex; the overall result, on the other hand, is simple: the fuel burns with oxygen to produce electricity and, as byproducts, carbon dioxide and water vapor (no unsaturated hydrocarbons and nitrogen oxides to form smog; cf. Section 2.1). Although one of the themes of this book is that we must depart from a reliance on fossil fuels because they will soon be exhausted, this does not apply to coal. A fuel cell which converted coal (or one of its products) *cleanly* to electric power (without ejecting SO_2 into the atmosphere) would be a valuable *interim* power source, perhaps necessary for the gap between the end of home-produced oil and the time at which solar and atomic reactors have been built in sufficient numbers. Of course, it would not be an acceptable solution for long because it would still continue to disturb the CO_2 balance.

Another approach is to use the petroleum products in a prereactor to form hydrogen, which is then fed to a hydrogen–oxygen cell. Because of the high electrochemical reactivity of hydrogen, the gain in the cell performance justifies the complication of the overall system. This is called "reforming" of the fuel. Reforming of hydrocarbons (oil, natural gas, and their derivatives) or that of carbon itself is done by the application of heat and steam, resulting in hydrogen, carbon dioxide, and carbon monoxide, the last of which can be utilized as a fuel itself. This application is only of temporary interest because it starts with fossil fuels.

Some other synthetic fuels, in addition to hydrogen, would be, for example, ammonia and hydrazine. More exotic fuels include sodium amalgam, solid metals, or glucose. Research is being carried out, for example, with the aim of producing an implantable fuel cell for heart pacemakers, which would take the "fuel" and the oxygen from the bloodstream of the body (such a power source would never need surgical removal for recharging). Fuel cells could be used for converting the sun's radiated energy into electricity. The chemical product of the fuel cell reaction can be broken up again into fuel and oxidizing agent under the influence of the sun's radiation. The coupling of such a photochemical reactor with the fuel cell would produce electricity without the need of a fuel source.

Engineering details and designs are just about as varied as the basic systems. Some operate at room temperature, others at high temperatures. Some use aqueous electrolytes, others ion

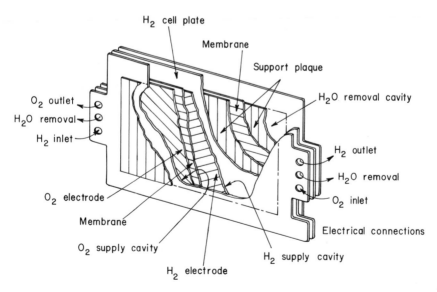

Fig. 5.2. A possible hydrogen–oxygen cell design.

exchange membranes, molten salts, or conductive solids. Almost invariably porous matrix electrodes are used to bring the reactants into contact with the electrolyte. An example of the design of a single cell is shown in Fig. 5.2. The practical system is more complicated and includes many cells in series, storage and feeding equipment for the reactant, and apparatus for the removal of the products, including waste heat; a possible arrangement is shown in Fig. 5.3.

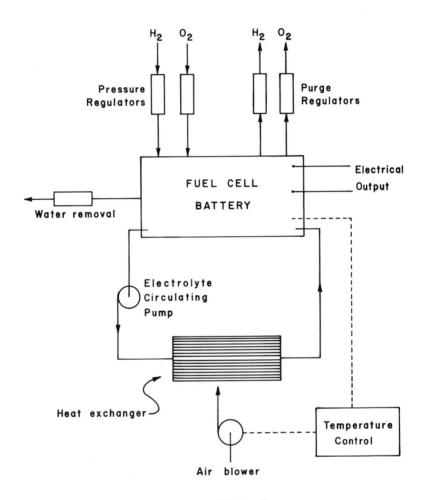

Fig. 5.3. An overall fuel cell system.

Table 5.1. Power Densities of Some Fuel Cell (and Other) Systems

Reactants	Power density, W/lb
Hydrogen–air (oxygen)	10–25
Methanol–air	25
Natural gas–air	10
Gasoline–air	15–35
Hydrazine–air	10–60
Aluminum–air	25–40
Lithium–chlorine	100–200
Internal combustion	*ca.* 400
Diesel	*ca.* 100

A summary of some fuel systems is shown in Table 5.1, indicating also power density. Energy density figures are not given since that depends very much on the fuel tank size and, therefore, is not a direct characteristic of the system. [Nevertheless, when *energy* density (not *power* density) *is* computed, it is much higher for fuel cells than for internal combustion engines.*]

5.3. Overall Characteristics of Fuel Cells, Examined with the Electric Car in Mind

How does the fuel cell measure up against the requirements set for the power source of a family car in Chapter 4? The figures of Table 5.1 indicate that *at present* none of the systems is adequate (with the exception of lithium–chlorine). Are further efforts in research and development needed to improve the performance justifiable? On ecological grounds, the answer is clearly *yes*. Some fuel cells would produce only harmless by-products (like water from the hydrogen–oxygen cell, and water and nitrogen from the ammonia and hydrazine cells) and obviously would be the final choice. Even the cells using hydrocarbon fuel are ahead of the internal combustion engine because of their high efficiency. Though they would produce carbon dioxide, it would be less per unit of energy produced than with the present methods of motive power generation.

* Some fuel cells would drive a car more than 1000 miles without refueling.

The inherent efficiency difference between the heat engines and the fuel cells is worth examining in more detail. The efficiency of all engines converting heat to mechanical work (which can then be converted to electrical energy) is limited by the Carnot theorem, giving the maximum possible efficiency as $(T_1 - T_2)/T_1$, where T_1 is the temperature of the working fluid entering the heat engine and T_2 is the temperature as it leaves the engine. The basic reasons behind the equation are complicated, but the equation and its consequences are straightforward. To get high efficiency, the end temperature should be near zero, or the working temperature very high. But a practical engine cannot reach near zero temperature since that would mean $-273°C$, and engines operate above ambient temperature to avoid the necessity of cooling at an expenditure of energy. There is also a limit for the highest working temperature because of construction material problems. As a result, the efficiency of the best heat engines is around 40%, the average, some 25%. *It is to be emphasized that these losses are inherent at given temperature limits, and cannot be engineered out by better design.* There are further "normal" efficiency losses due to friction, heat losses, etc., which can be minimized, but not reduced to zero. As a result, the overall efficiency of car engines is 15–25%.

One of the advantages of fuel cells is the absence of Carnot limitations, since the chemical energy of the fuels is not going through the heat–mechanical energy–electricity chain, but is *directly* converted into electricity. There are, of course, some efficiency losses also in the fuel cell, but they are about one-third those of internal combustion engines, resulting in an overall efficiency range of 60–80%. This is the reason that, while we still do burn fossil fuels, fuel cells would be ecologically better than the internal combustion engine. A fuel cell operating at 75% efficiency, compared to a heat engine of 25%, will produce only one-third of the carbon dioxide for the same amount of power generated. While a long-lived affluent world will only be possible after carbon dioxide has been eliminated, the reduced emission of CO_2 achieved by a fuel cell burning fossil fuels could be a valuable solution for some years.

5.4. Further Uses of Fuel Cells

In this chapter, as in the preceding one, the transportational uses of electrochemical power sources are stressed, with special emphasis on the family car. Many other possible uses of fuel cells are, of course, also envisioned.

There are applications where fuel cells in their present form are already competitive. In space exploration they are more than competitive, and unique in value because no other method allows so much energy to be produced for the same amount of weight. In the moon missions, hydrogen–oxygen fuel cells were used. Fuel cell power stations for remote locations are commercially available; in this case, the higher costs are outweighed by the savings in fuel transportation costs due to the high efficiency.

In mobile applications, there are many areas where the requirements are not as stringent as for a family car: small commuter cars, delivery trucks, forklifts and other vehicles operated in close space (the internal combustion engine is already outlawed in many states from use in closed areas), tractors, rail transportation, etc. Marine applications are also promising for smaller ships, where nuclear plants will not be economical, and in the case of submarines, where the quiet operation and longer range than are possible with batteries are added advantages. Mobile recreational uses (boating, camping, scuba diving) should be mentioned, for here the weight of fuel to be carried may be significant.

Home electricity generation is one of the most important possibilities. A set of fuel cells in the basement, generating all the electricity needs of the household, would have many advantages. While fossil fuels are still being used, such an arrangement would better utilize the dwindling reserves of natural gas and oil, and at the same time reduce the carbon dioxide emission to the atmosphere, because of the higher efficiency of energy conversion.* Furthermore, an underground pipeline system for

* Precisely such a fuel cell system, using natural gas, is undergoing on-site

fuel distribution is esthetically preferable to overhead cables. Here again, the fossil fuel cells can only be a temporary solution. In an all-electric economy, as will be discussed in the next chapter, synthetic fuels will compete with transported electricity and the household fuel cell may become one of the central points of the energy distribution system.

5.5. The Most Developed Fuel Cell System

The hydrogen–oxygen (air) cell is the one system which attracts the most attention among scientists and engineers alike. As was mentioned before, this is the system which was developed to the point of practical usefulness. Both the Gemini and Apollo space programs have used these fuel cells for electrical energy generation (cf. Fig. 5.4), and they are reliable.

The chemistry of the cell is simple; it has been discussed in Section 5.1. The engineering approaches to the realization of the cell will be briefly sketched here only to show the multitude of possibilities for further development.

Most of the work has been carried out with the low-temperature cells (below the boiling point of water), the application of which would, quite understandably, be the most convenient. Usually highly basic electrolytes are used, though lately acidic electrolytes have also gained favor. The electrodes are carbon or nickel, and because of the low temperature, require a noble metal catalyst coating. In one special cell, the aqueous electrolyte was replaced by an ionically conducting plastic sheet (called an ion exchange resin membrane). The power density of these cells is typically in the 10–20 W/lb range (cf. Table 5.1).

In the higher-temperature cells (which are sometimes operated at higher than atmospheric pressures), the solutions can be more concentrated—sometimes molten salts are used. The electrodes are less expensive since the high temperature accelerates the reactions and noble metal catalysts are not needed.

---

test procedures during 1973, and will be built in 1974 into a 50 megawatt producer. It is funded by many partners in the United States gas industry.

Fig. 5.4. Final assembly of fuel cells for the Apollo spacecraft (Photo courtesy Pratt and Whitney Aircraft).

Table 5.2. Some Possible Fuels

Fuel	Energy content at 80% efficiency, kWH/gal	Estimated price, ¢/gal	Power cost, ¢/kWH
Hydrogen	0.3	18	60
Ammonia	0.36	20	56
Hydrazine	0.59	590	1000
Methanol	0.56	10	18
Diesel fuel	1.18	20	17

On the other hand, general construction material problems are increased because of the more corrosive nature of the high-temperature electrolytes. Power densities of about 10–15 W/lb have been achieved. One cell operating at *ca.* 1000°C uses solid ceramic material as electrolyte, onto which the electrodes are deposited as coatings. The cell is compact, producing 20–45 W/lb.*

5.6. Objective Factors Which Affect the Progress of Fuel Cells

(i) Fuel Costs

In attempting to devise new fuel cells, there is no point (if the powering of transportation is the aim) in going outside fuels which are cheap. This means that diesel oil, methanol, hydrogen, and perhaps hydrazine and ammonia should be considered. The cost of each of these per kilowatt-hour of electricity contained in them is shown in Table 5.2.

---

* These *power* density figures show that this factor is relatively poor for fuel cells developed up to the present. They should be compared with gasoline engines, at about 400 W/lb, or diesel engines at about 100 W/lb. However, in terms of *energy* density, for use, without refueling of about 12 hours, the fuel cell has about 400 WH/lb, and the gasoline–air system only about 70 WH/lb. Fuel cells are the lightest of all energy storers for times of use up to about 100 hours. The differences of the characteristics for power and energy densities are much to be noted.

Among these fuels, the use of hydrogen, hydrazine, and ammonia is environmentally preferable to the use of diesel oil or methanol. Thus, if we used the latter, we should produce carbon dioxide as the final product! This would seem to be no great improvement on the internal combustion engine ecologically. However, it *would* be an improvement, because, owing to the greater efficiency of the fuel cell, much less CO_2 would be produced per mile of travel, as discussed before.

(ii) Electrocatalysis

In electrochemical power sources, the rate at which an electrode reaction occurs is connected, as we show in Appendix II, to the development of an overpotential, which depends on the rate constant of the electrochemical reaction. The larger this rate constant, the smaller will be the overpotential which we need to achieve a certain reaction rate, i.e., develop a certain amount of power from our fuel cell. But the smaller the overpotential, the smaller the loss in efficiency and the cheaper the electricity.

Hence, one of the most important matters in the development of fuel cells is to get the electrocatalysis *high*. Hypothetically, if we could get an "exceedingly good" electrocatalyst, then we would have to develop negligible overpotential to make the fuel cell deliver high power.

What would that mean? The efficiency of the fuel cell is proportional to the ratio of the available cell voltage (theoretical voltage diminished by the overpotential) to the theoretical voltage, so that the efficiency is diminished by an increase in the overpotential.

The key to the development of fuel cells is, therefore, a knowledge of the relatively new subject (born in 1963) of electrocatalysis. The importance of this subject can be shown with a few examples. The oxidation rate of methanol (a possible fuel) can be increased about a million times by the use of good electrocatalysts, and such a difference may well make the difference between a useless and a practical system. Unfortunately, good catalysts (usually noble metals) are expensive and for a fuel cell the catalyst alone may cost several hundred dollars for each kilowatt of power capacity. The development of good *cheap* catalysts is, therefore, vital for the development of fuel cells. Alternatively, it may be possible to find ways whereby noble metals such as platinum—good catalysts but too expensive—are used in quantities so tiny (i.e., added as "bits" on the surface of cheap substrates) that they become sufficiently cheap per kilowatt of power.

There is a particular reaction relevant to the study of fuel cells. It is the catalysis of the reduction of oxygen from air. Electrochemically, this reaction follows the equation

$$O_2 + 4H^+ + 4e \rightarrow 2H_2O$$

This reaction is of great significance because it is one of the reactions in nearly all fuel cells. Furthermore, oxygen reduction (probably electrochemical) is an important reaction in bio-energetics. It is this reaction of oxygen reduction which often causes the maximum overpotential in fuel cells. Were its mechanism understood, it would probably be easier to get on with finding a cheap electrocatalyst, and therefore improve the performance of fuel cells. Thus, hemoglobin in biological systems is an effective oxygen reduction catalyst. It would be interesting to know why, and to attempt to apply the lesson to develop a catalyst in a fuel cell.

(iii) Transport of Materials to the Interface

The development of suitable electrodes (we could call them "catalyst supports") enabling the fuel (for example, hydrogen) to gain easy access to the catalyst is not simple. Thus, catalysts in fuel cells are laid down in porous electrodes, shown in Fig. 5.5.

This porous structure serves not only to bring the fuel, the electrolyte, and the electrode-catalyst into contact, but also to serve as a container for the solution and gas (the liquid should not flow out, neither should the gas bubble through). However, in addition to the tightness requirement (for the container action), there is a contradictory requirement of openness for ease of fuel transport to the catalyst. The problem is complex and cannot

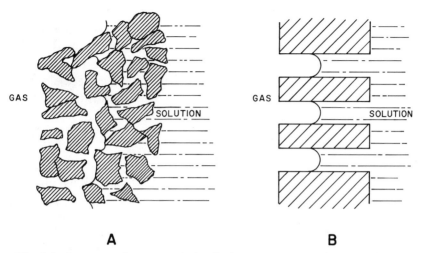

Fig. 5.5. Structure of a porous fuel cell electrode: (A) actual; (B) idealized. The catalyst is not shown. It often consists of small pellets and microcrystals on the pores.

be treated here in detail (however, see Appendix III). Some successful solutions have already been obtained, but there is much engineering work to be done in the development of suitable porous electrode structures.

(iv) Political Factors Which Retard the Development of Batteries and Fuel Cells

The rate of development of our future depends upon political and sociological factors rather than the scientific and engineering ones. This is particularly so in the development of batteries and fuel cells, in particular because of the gigantic economic consequences of their successful use in transportation.

Thus, were research to lead to the development of a good high-energy-density battery, which would give energy per unit weight comparable with that of an internal combustion engine, there would be no scientific impediment to the development of an ecologically acceptable, electrochemically powered transportation system. However, such a change would have a dislocating effect on the automotive and oil industries. These are the largest industrial complexes in existence—together with the closely connected steel industry, they are the origin of some half of the economic activity of the United States. The effect on the oil industry of a successful nonpolluting electric car is obvious. The car industries would be negatively affected by the necessity of radical and far-reaching changes in the manufacturing process, amounting to a rebuilding of the industry. The steel industry would be fatally wounded by electric cars, which are likely to be small and built of plastic or lightweight alloys.

One does not need to impute Machiavellian thoughts to the directors of the mammoth international corporations about electrochemistry and its development: *their* job is to make a good profit. Furthermore, considering the effect on the national economy, the evolution away from oil and the internal combustion engine car must be carried out over a suitable period, at least one decade, in order to be economically acceptable. The danger is more that the change will be made too slowly to be *ecologically* acceptable.

5.7. What Electrochemical Power Sources Will Run Cars in the Near Future?

The answer to this question grinds on the rocks of the uncertainties of government policy in developing abundant electricity, and to what extent there is a prospect of a hydrogen economy (see Chapter 6).

Sufficient rapid development of breeders or satisfactory solar energy would send us toward batteries; lack in the development of abundant clean electricity would send us toward fuel cells (where the main direction would be toward a noncarbonaceous fuel, sufficiently cheap, and having acceptable emission characteristics, but *not* needing abundant cheap electricity to make).*

---

* Hitherto (1973), United States government policy has not shown any consciousness of the lateness of the hour in beginning the gigantic research and development program needed to replace fossil fuels. It relies, at present, on the development of the even more lethally polluting atomic energy.

The second point is that many published discussions given in other sources mislead for the following reason: asked what nonpolluting electrochemical power sources would replace the internal combustion engine, most automotive engineers think of the characteristics shown by the internal combustion engine as the test point, *they do not think ecologically*. They orient to supply what they say the customer wants (in reality what he has been conditioned by advertisements to want)—behemoths, with monstrous acceleration and vast spare power, which we now, so astoundingly, use for our transportation (often a 5000-lb vehicles for the transport of one 150-lb individual). Present (stimulated) customer wish and optimization of the economics *in the short term* power the response, not ecological necessity and long-term communal need.

If the objective is a rational means for intraurban transportation at speeds up to 55 mph, with a range without charging in the region of 100 miles, then the technology needed to produce it is already here. One could have a car which would have the characteristics of the experimental electric car of Union Carbide Company, which uses a hybrid power system combining the high power density of a bank of batteries with the high energy density of the fuel cell. The actual system uses lead–acid batteries with 50 W/lb and 11 WH/lb characteristics, and a hydrogen–air fuel cell with 17 W/lb and 100 WH/lb power and energy densities, respectively.* The car has a range of 200 miles, "fills up" in 3 minutes, and has a top speed of 55 mph and acceleration characteristics comparable to today's compact cars with internal combustion engines. A hybrid type of design of this kind could be developed by replacing the lead–acid battery by a zinc–air battery, or perhaps by some other kind of metal–air cell for acceleration.

A further and better development—assuming the availability of abundant electricity, but no hydrogen economy— would be that of high-temperature cells. Here, for example, with lithium–chlorine, there does appear to be a probability

---

* The complementary characteristics of batteries and fuel cells (batteries, high *power* density; and fuel cells, high *energy* density) are important to note.

of obtaining a power source in which the power-to-weight ratio (taking everything into account and looking at the whole vehicle) would be *comparable* to (perhaps half as good as) that of the polluting internal combustion engine. There would, of course, be very great advantages in other directions, looking away from the power per unit weight, above all complete absence of pollution, little noise, and no vibration. However, there is some technology still to be developed, in particular the stability of the cells, which at present are too unreliable.

Lastly, when hydrogen becomes the universal fuel (Chapter 6), we shall run everything on hydrogen–air fuel cells, though we might still want to have booster batteries for rapid acceleration (the main power in a vehicle power plant is switched on for acceleration and hill climbing, only *ca.*30 % is required for constant-speed cruising). This is what makes reasonable the prospect of a hybrid system—a fuel cell which would develop power for long-range cruising working in conjunction with a battery which would give boosting power for acceleration and hill weight. (It would be recharged during steady driving by the fuel cell.)

5.8. Illusions with Respect to the Electric Automobile

The following are frequently stated misapprehensions.

(i) *Recharging would be slow.* But one can rent charged batteries and not have to wait for recharging. If fuel cells were used, one would simply "fill up" as now.

(ii) *Electrochemically powered transportation is "dirty" too, because the extra electricity must be made somewhere and that would cause pollution.* Of course, there would be no object whatsoever in having an electrochemically powered transportation system, if one made the extra electricity which would be needed in a dirty way, i.e., from coal or oil. The assumption is that it would be atomic or solar, but obviously, absolutely clean.

(iii) *Electrochemically powered transportation would be expensive.* It will be cheaper to run (see Table 5.3).

Table 5.3. Comparative Cost of Fuels

Fuel	Price	Cents per horsepower-hour	Cents per mile for a typical (20 mpg) car
n-Octane	25 ¢/gal	3.5	1.25
Hydrogen	30 ¢/lb	1.9	0.68

Note: 15% efficiency was assumed for *n*-octane (internal combustion); 65% efficiency was assumed for hydrogen (fuel cell).

(iv) *Much research is being done to develop new types of cheap batteries and fuel cells for cars.* Many people believe this is the case. At the time of writing (1973), *hardly any work was being funded in the United States in this direction,* outside the work in the automotive companies, which would be so badly hurt were their work to succeed.

United States Government support of research in clean electrochemical power sources was reduced by about 75 % between 1969 and 1970, about the period in which the press reflected a considerable expression of public desire to leave the internal combustion engine and get into a new and clean power source.

The NASA development of lunar travel cost approximately $25 billion. What we spend per year on the development of an electrochemical power source for clean transportation is (outside the defensive research of the automotive companies!) perphaps about $1 million. An appropriate level of spending would seem to be at least that which NASA spent on developing fuel cells for the lunar transports, which was in the vicinity of $10 million per year, over about 8 years. The development of a high-energy-density nonpolluting electrochemical power source for cars is probably a more difficult task than the development of an electrochemical energy source for space vehicles. But much of our future depends upon the development of clean power sources whereby we shall have the benefits of transportation without the negative ecological implications of not running them electrically. The development of a lightweight, high-powered battery is a worthy national priority item for most countries. We have perhaps 20 years before the possession of such a battery—or other electrochemical device—becomes an utter necessity. That is approximately the time it would probably take to develop one, starting in the mid 1970's.

(v) *Electric cars will have a poor performance.* Made in 1973, they would tend to have a performance of up to 60 mph with a range of *ca.* 100 miles. Made in 1985, and assuming there is funding of research in electrochemical power sources in the meantime at a level corresponding, say, to that used for the development of the fuel cells in space vehicles, they could have the same performance as gasoline-driven cars now.

(vi) *We would need a large amount of new power generation facilities (of course, solar or atomic) to produce the increased amount of electricity needed to run electric cars.* In the case of fuel cells this, of course, does not apply, and it is not correct for the case where cars would be fully powered by rechargeable batteries. The use of electricity declines sharply overnight, and the generating facilities needed to meet the peak demands during the day are underutilized at night. These off-peak hours could be used to generate the electricity needed for the electric cars. If one charges the batteries at home, the most convenient time to do it will be overnight. If one rents charged batteries, the discharged ones can again be recharged overnight. From the projected growth of power generating facilities from now to the year 2000, and the estimated car population growth, it can be shown that off-peak electricity will be able to supply all the energy needed for electric transportation without the necessity for building new power plants for that purpose.

5.9. Summary of the Situation with Regard to Electrochemical Power Sources

A point to remember is that it is not a question of whether an electrochemically driven transport would be better than one driven by an internal combustion engine. This is the way the matter has been judged in the recent past. That is no longer the question; we *must* have, by 1990, mass production of non-polluting and non-CO_2-producing vehicles. No other suggestions for nonpolluting vehicles have been made, except the electric car.

The question is, how will it be powered? There are three possibilities:

(i)   Overhead cable, third rail, etc., with the limitation in mobility that this will give.
(ii)   Batteries.
(iii) Fuel cells.

The development of the latter would seem to be advantageous not only to attain the considerable goal of powering cars in a nonpolluting way, but because we shall have to power also ships, and eventually planes, by electrical means, and these cannot be powered by a system needing "overhead cables."

One should not forget the possibilities of research on other *principles* for electric energy storage. One of these might be storage in condensers, where problems at the present time have to do with the potential at which dielectric breakdown occurs, also a subject which could benefit greatly from research.

CHAPTER 6

The Hydrogen Economy

6.1. The Inevitability of an All-Electric Economy

Before describing the characteristics of an economy in which hydrogen is the medium of energy, let us summarize briefly the reasons leading to it. As pointed out above, man's energy source will have to be changed during the next two or three decades. Such a change is forced upon us by irrevocable events: the increasing atmospheric pollution caused by the burning of oil, natural gas, and coal to obtain energy, and the diminishing reserves of oil and natural gas. Because of the necessity of carrying out this change in only two to three decades, we shall have to accept fission reactors at first (presently, the "conventional" ones, soon the breeders); and, if one looks somewhat further ahead in time—two to three decades with ordinary funding—one can hopefully add utilization of the sun's radiant energy and, eventually, perhaps, fusion reactors.

How will these new types of power sources affect the way we use our energy? The main forms of energy used by man are heat and mechanical work. Atomic reactors produce heat directly, and this can then be converted to mechanical energy. But one cannot put an atomic reactor in the basement and shovel uranium into it to heat the house, neither can a reactor be put into a car to produce the mechanical energy necessary for personal transportation. Further, it cannot be put in the back-yard of the local manufacturing company, as some factories now produce their own power by steam-electric generators. By their nature, the new power generating plants—atomic or solar—will be huge installations, producing very large amounts of energy, and occupying large areas of land. They will be complex installations, requiring extensive auxiliary and con-trol equipment and, for protection from radiation, heavy shielding. By increasing the size of the reactor, and so its power output, the needed auxiliary equipment and shielding are increased but little. Therefore, the larger the reactor, the lower the cost per unit of energy. Present reactors are in the hundreds of megawatts range, but reactors now being built have capac-ities of several thousand megawatts. Because of safety precau-tions and the huge cooling water requirements, these stations will have to be located in the oceans, on man-made floating islands, tens of miles offshore.* The power from these stations will have to be transported to the user, who may be a thousand miles away inland. For the solar reactors the situation will be even more pronounced, for the sun's radiation is more available in the Sahara Desert, Saudi Arabia, and above all Australia than elsewhere.

Thus, a principal characteristic of the new power sources will be that the power will have to be transported over large distances (>500 miles) from the source to the user. How can this be accomplished? Heat or mechanical energy cannot be transported. The present way to accomplish the job would be to turn the energy into electricity and carry it in cables to the user. Future power stations (atomic or solar) will give electrici-ty directly.

---

* The energy islands must not be too near the shore because of the necessity of being in sufficiently deep water to make the pumping of cold water from deep levels possible. A problem of defense against military attack would arise. The solution of building the giant reactors entirely underground does not seem acceptable because of the heat production and the necessity of vast amounts of cooling water. A better solution seems to lie in the mutual nature of the vulnerability.

6.2. Transmission of Energy over Long Distances

Transmission and distribution costs of electricity form a major contribution to its price. The fraction of this contribution increases with the distance from the source, particularly when the cost of energy at its source becomes less than at present, as one would expect it to with the future sources. Thus, although the production cost at source will be reduced with the large atomic or solar plants, this could be counterbalanced by transmission costs unless a less expensive way of accomplishing energy transport is found. At present, electricity is not transmitted in cables over thousands of miles, but fuel (e.g., oil in pipes) is carried to the local plants. This suggests an alternative for the future. A "synthetic fuel" can be produced at the site of the power generating plant and transported in pipelines, tankcars, or tankers to the point of use, where it is converted to electricity or some other form of energy. Hydrogen has been suggested as such a fuel. The transportation costs of electricity in high-voltage cables and that of hydrogen in a pipeline are compared in Fig. 6.1. For long distances, the savings are significant using hydrogen as the medium of energy. In addition, the hydrogen-containing pipes would be underground, so that unsightly cables would be absent. As far as energy transmission is concerned, therefore, this method offers an advantageous alternative to the transmission of electricity in a grid of cables. Of course, the problems of converting the energy into a synthetic fuel, and its reconversion into energy, will have to be considered: they will be discussed in later sections.

Other synthetic fuels would be possible. A carbon-containing material, for example, methanol, can be eliminated since its use would continue to pollute the atmosphere with carbon dioxide.* Only fuels producing no polluting side-

---

* Remarkably, Jersey Enterprises, a subsidiary of the Standard Oil Company of New Jersey, is involved in an intensive research program in collaboration with the French company, Alsthom, to produce fuel cells which run on

products can be considered. This is the reason why hydrogen is a likely choice since it would produce only water when used as a fuel. Even nitrogen-containing fuels, such as ammonia or hydrazine, could be troublesome from the pollutional standpoint. In a fuel cell, they would produce only nitrogen and water vapor as exhaust gases, but in the case of direct burning (a likely use for heating purposes), the emission of nitrogen oxides would be a problem.

The advantage of the hydrogen economy due to the elimination of overhead cables must not be overlooked. At present, there are about 300,000 miles of transmission cables in the United States; half a million miles is predicted for 1990. This means about 11,000 square miles of right-of-way land use for the power transmission (larger than the area of the state of Maryland). Pipelines also need right-of-way land, but the fact that they are invisible and do not spoil the landscape has an advantage, though it may be hard to express it in dollars and cents. The ever-increasing trend to take environmental effects into consideration will be an impetus for an underground pipeline system against the overhead cables. Electric cables can, in principle, also be placed underground, but only at an extremely high relative cost, as indicated in Fig. 6.1.

6.3. Production Methods for Hydrogen

How can the energy produced by a nuclear power plant be used to make hydrogen? The presently preferred hydrogen production methods involve the use of carbonaceous materials, being mainly the steam reforming of natural gas, hydrocarbons, or coal (coke). None of these processes is acceptable for the future: they consume fossil fuels and produce carbon

---

methanol, and could be the basis of an electrochemical power source for cars. Although such an approach may be economically viable, it is ecologically half as bad as the use of fossil fuels, as far as the continued injection of CO_2 into the atmosphere is concerned. It would have perhaps twice the efficiency of internal combustion engines, i.e., produce half as much CO_2 per mile.

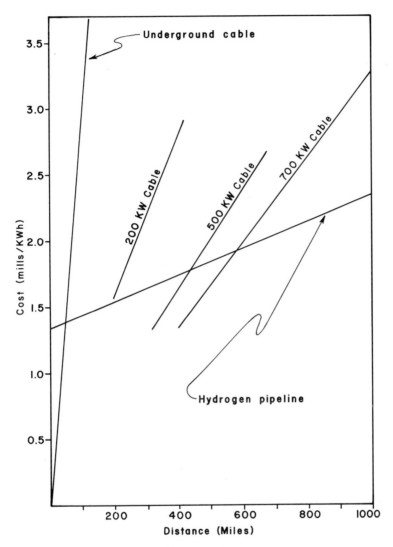

Fig. 6.1. Comparative costs of power transmission.

dioxide and other pollutants. There is, however, another well-known hydrogen production technique: the electrochemical production of hydrogen from water. This is a well developed (engineered) process used presently in many locations where the price of electricity is low.

Water electrolysis is chemically simple: it is the reverse of the hydrogen—oxygen fuel cell (Sections 5.1 and 5.5). At one electrode, hydrogen gas is produced through the effect of the electric current, and oxygen gas is given off at the other electrode. That is, the spontaneous reaction of the fuel cell (which *produces* electricity) is pushed in the reverse direction, with consumption of electrical energy. The individual reactions at electrodes are complex (Appendix II), but the overall reaction is simple, i.e.,

$$2H_2O \rightarrow 2H_2 + O_2$$

A cell for producing hydrogen is shown in Fig. 6.2. A porous diaphragm is used between the electrodes which allows the the current to flow in the solution between the electrodes but eliminates mixing of the two gaseous products. Though the reaction is the decomposition of water, an electrolyte (e.g., sodium hydroxide) is dissolved in the water to increase its conductivity. Many different and well tried cell designs are commercially available, and new designs (some of them spin-offs from fuel cell research) are constantly suggested. The overall efficiency of the present process is 60–65%, and efficiencies of the newly developing cells are projected to be between 70 and 80%.

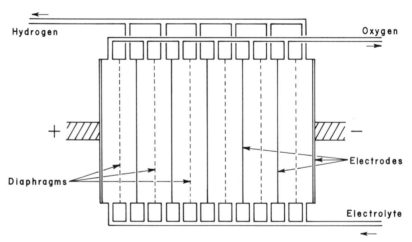

Fig. 6.2. Schematic of a water electrolyzer.

Some other possible processes for the production of hydrogen at a nuclear plant site are the decomposition of water by nuclear radiation or by the use of heat to bring about chemical reactions in which metal oxides are used as intermediates between water and the final product, H_2. One disadvantage of such methods is that they involve high temperatures (> 500°C), and this may increase the materials cost. None of these processes is well known or developed enough at this stage to be a serious competitor for the electrolytic production, though this might change in the future.

The raw material situation must be stressed. For hydrogen production only water is needed as raw material with the above discussed processes, and that is something which will be in plentiful supply around the reactors if they are placed on floating islands in the ocean.

Correspondingly, when solar farms are built, similar arguments apply. There is less point to put solar reactors on the sea because their problems of heat disposal are less. If they are not on the sea, water for the electrolysis would have to be available near them. In Australia, for example, there are extensive aquifers at depths of about 2000 ft, and a number of these have plentiful water at temperatures of about 100–150 C, good raw material for electrolysis.

6.4. Reconversion of Hydrogen to Energy at the Point of Use

To what forms of energy do we want to reconvert the hydrogen fuel at the point of use, and how can this be done? The main forms of energy utilized are electricity, heat, and mechanical work. Let us examine each of them separately.

Electricity, as such, is used at present largely for lighting and communications. It is also used indirectly (and it will increasingly be used in the future for ecological regions) in many industrial processes: waste purification, material recycling, manufacturing; electrochemical aspects of these will be presented in the following chapters. The way hydrogen can be used to produce electricity is by means of the fuel cell, and that was explored in the preceding chapter. Manufacturing plants and housing complexes will have their fuel cell stations where electricity will be produced from hydrogen and air and directed locally to the site of use. Individual homes may either have 10-kW fuel cells in the basement, or larger fuel cell installations may be located at power substations from which electricity is distributed by underground cables for short distances.

Heating uses of hydrogen may be direct ones. Hydrogen can be burned with air, like today's fossil fuels, to produce heat; the only by-product of this process will be water vapor, no soot, fumes, or other pollution. The heat content of hydrogen per unit volume is lower than that of the presently used fuel gases (because of its low density), so that a redesign of heating equipment would be necessary (hydrogen at present is already used for heating applications though usually mixed with hydrocarbons). If more convenient, hydrogen could be converted to electricity in fuel cells, and the electricity used for heating. The efficiencies are about the same. The fuel cell converts hydrogen into electrical energy with an efficiency of about 60–70%, and the conversion of this energy to heat is nearly 100% efficient; direct heaters work in the 60–80% efficiency range. The situation is different with today's electrical heating, which is ecologically unwise. Thus, electricity today is generated at 30–40% efficiency, while 60–70% of the fuel's energy is vented through the powerhouse chimney. Relatively, present electric heating uses twice as much energy per unit of heat available at a site as would electric heating run on hydrogen-produced electricity.

For the generation of mechanical work (transportation on the ground, in water, and possibly in the air*) the fuel cell

---

* However, the use of fuel cells to run planes is beyond the presently foreseen technology: fuel cells are very light energy transducers if much energy is needed and *power* density is less important. In aircraft, very high (*ca.* 10 kW/lb) *power* is needed, and the power per unit weight of fuel cells cannot at present be anticipated as becoming comparable with that of combustion engines.

will play a central role. The most efficient way to use hydrogen to obtain mechanical work will be by converting it first to electricity in a fuel cell and then using it in lightweight motors. The combination of fuel cell–electric motor will have a higher overall efficiency (fuel cells, 60–70%, motor, >90%) than a heat engine with a 15–20% efficiency. Hydrogen could be used in engines similar to today's internal or external combustion engines, but that would involve a waste of four-fifth of the energy as heat and thus lead to the necessity of fuel use twice as large as with the electrical route. Also, use of hydrogen and air in combustion engines to obtain mechanical power would produce NO_x as a pollutant, whereas the only by-product of the fuel cell is distilled water.

6.5. Some Miscellaneous Aspects of the Hydrogen Economy

(i) Nonenergy Uses of Hydrogen

Presently there is about 6000 billions of cubic feet of hydrogen produced in the world, practically all of which is used in the chemical industry. It is an important raw material. Ammonia (and from this fertilizers), hydrogenation of vegetable and animal oils, and hydrocarbon processing are some of the most important uses presently. With abundant, low-cost hydrogen available—the all-electric economy making it and and using it as a synthetic fuel—many processes using hydrogen which are at present marginal will become economical, increasing its use as a chemical raw material. Its ecological advantages will make its use more widespread. Consider the possibility of the reduction of iron ores directly by hydrogen. In the reaction of iron oxides and hydrogen, the only products will be the iron and water vapor. The steel plant, a major polluter today, could become clean. Coal would be conserved for nonpolluting uses (presently one-fourth of our total coal production is used by the steel industry). It could be used, for example, with hydrogen for the production of synthetic hydrocarbons, supplying raw material for the plastic and chemical industries, and yield chemicals which could be the basis, along with nitrogen from the air and enzymes, in the synthesis of protein foods. Iron was mentioned as an example; hydrogen could be used similarly to eliminate pollution in many other metallurgical processes.

(ii) By-products of Hydrogen Production

Hydrogen production will give some useful by-products. Decomposition of water will yield oxygen in addition to hydrogen. Oxygen presently is produced by the liquefaction of air and is widely used as a chemical raw material, for example, in the processing of steel. The amounts which would become available as a by-product of the hydrogen economy are enormous (half a cubic foot for each cubic foot of hydrogen).

Another by-product of the electrolytic hydrogen production is heavy water (deuterium oxide).* There is a difference in the rates at which hydrogen and deuterium are evolved from water. The hydrogen is evolved preferentially during the electrolysis, resulting in an enrichment in the heavy water content of the residual water. This might become an important by-product since deuterium is one of the raw materials which may one day be used in fusion reactors (*cf.* Section 3.4). In any case, the sea would have to be "mined" (electrolyzed) to get the deuterium. In the hydrogen economy, therefore, the conversion of nuclear energy into synthetic fuel can be combined with the mining of fuel for possible fusion reactors.

(iii) The By-product of Energy Production from Hydrogen

We have stressed the nonpolluting aspect of hydrogen as a fuel, giving only water as a by-product. This by-product may turn out to be important in its own right. With the growth of the population, clean, potable water is getting into short supply. It would be easy to condense the water vapor from the

---

* Deuterium is a heavy isotope of hydrogen, containing in its nucleus a proton and a neutron, while the hydrogen nucleus is only a proton.

exhaust of fuel cells. The fuel cells used in the Gemini space flight produced one pint of drinking water for each kilowatt-hour of electricity. If one assumes that, in the entirely electric, nonpolluting economy, a household would use 80 WH of electricity daily (the present use is about 35 WH), ten gallons per day of pure water could be obtained as a by-product. While this is less than the total needs of the family, it should be enough for internal consumption (drinking water, cooking). It may be that, in a hydrogen economy, one will have two kinds of faucets in the house, one for potable water, and one for cleaning purposes only. If the potable water is supplied by the fuel cells, the municipal water recycling system will be able to produce clean (but not drinkable) water at a reduced cost. This cleaning water will still have to meet sanitary standards (bacteria count, etc.), but it could have a much higher saline content than water for drinking and cooking. Every lavatory flush costs several gallons. For such purposes, even sea water would be acceptable.

(iv) Energy Storage Aspects of the Hydrogen Economy

Up to now, we have considered the transportation of energy only; another important aspect of energy management is storage. Hydrogen offers a possibility for this too. Once the electricity is converted into the hydrogen synthetic fuel, it can be stored for any amount of time before being reconverted to electricity again. The storage capability is versatile; it can be done at the point of energy production (e.g., in cryogenic containers in the sea under the floating energy islands), at any distribution centers or substations, or at the point of use, whichever is the most economical and convenient. With such storage facilities, the production of electricity at an atomic power plant could be carried out with a continuous uniform load 24 hours a day, seven days a week, the most efficient way of operation. The use and production patterns of energy could be separated from each other by a buffer hydrogen storage system between the power plant and the consumer. Limitless storage space for cryogenic hydrogen exists under the sea, at depths where the temperature is significantly below the surface temperature.

An interesting prospect is the use of the pressure gradient in the sea for aiding the liquefaction and storage of hydrogen. In order to liquefy the gas, it must be compressed. By arranging electrolysis at considerable depth, the hydrogen could be produced at a high pressure, thus diminishing the mechanical work in the liquefaction.*

6.6. Safety Aspects of the Hydrogen Economy

Safety considerations are always a topic when hydrogen usage is under consideration. One is taught at school that hydrogen is dangerous material, because it can form explosive mixtures with air, has no odor (detectable only by instruments), and leaks easily because of its low viscosity (the factor which makes it easy to pump). However, hydrogen has been used in industry for many decades and adequately safe handling and storage procedures have been developed, which can form a basis of safe manipulation of hydrogen as a widely used synthetic fuel. Hydrogen, properly handled, is no more dangerous than other materials commonly in use now. Today's fuels (e.g., natural gas or gasoline) are dangerous too, and can cause explosions, fires, and accidents. In some ways they are more dangerous than hydrogen, because their vapors are heavier and hence they remain at low level, subject to initiation of an explosion by sparking. Through safe handling, these occurrences are kept to a minimum with present fuels. The same can be accomplished with hydrogen, given proper engineering attention. For many years, tankcars containing liquid hydrogen have passed on trains through towns and tunnels without difficulty.

---

* The production of high-pressure gas requires more electrical energy than that at atmospheric pressure (both the equilibrium potential and the polarization losses are pressure dependent), but present industrial practice shows that high-pressure water electrolysis is more economical.

6.7. Summary

We have given evidence which indicates that, through a combination of the exhaustion of oil and natural gas, and the polluting nature of burning coal, there must be an all-electric economy in being in the early years of the next century. With a needed research and development time of 10–15 years, and a changeover period of 15–25 years, we should be well into the R & D stage for this change. The change will be a much bigger one than that of our former change from coal to oil, or the earlier ones from wind power to coal, and not to be greeted with jubilation among the population, for its only overt gain will be the gradual fading away of pollution, and it may cause some disruption and very much rebuilding. However, it is completely inevitable and the only big questions are when solar energy and nuclear fusion processes will be introduced.

The desirability of a "synthetic fuel" to be used parallel with electricity has also been explained, and it was shown that hydrogen is strong contender for such a use. A summary of such a hydrogen economy will be given here. Atomic or solar reactors, located on man-made offshore islands in the oceans, or suitable desert areas on shore, would produce abundant and low-cost electricity. The electrical energy is used at the plant site to electrolyze water with the production of hydrogen as a synthetic fuel; the deuterium by-product of the process may be the basic fuel of fusion reactors produced in the next century. The hydrogen transported in pipelines to convenient storage locations, from where it is distributed to the consumer. Because of this buffer storage, the power plants can operate at a uniform load, independently of irregularities of the energy use pattern, resulting in lower power costs. The consumer receives the hydrogen in pipelines and reconverts it to electricity in hydrogen–air fuel cells, which in addition to supplying the household's energy, will also produce its potable water. This can be accomplished with no environmental pollution at all if solar radiation is the origin of the energy. Heat pollution from reactors would be dissipated in the deep ocean, the hydrogen would be transported in pipelines, and the only by-product of electricity production at the point of use would be clean water which can be collected and used. Much of the basic knowledge needed for the engineering of such a system is available now. The realization of such an energy system (the end of the utility of which cannot be foreseen) is within grasp. The time at which it will come about, nearer to 25 years hence or nearer to 50, is largely a matter of the spread of information and understanding, and the subsequent exertion of democratic forces on the legislators (in competition with the oppositely directed plutocratic ones).

CHAPTER 7

Electrochemical Waste Treatment

7.1. The Waste Problem

Man has always generated wastes, materials which were either by-products of his activities, for which he could not find any use, or products which have reached the end of their useful life. While this has been going on throughout the ages, no problem was caused until recent times because of nature's own waste treatment processes: dispersion, dilution, and degradation. The smoke of the caveman's fire (even the smoke of last century's industry) was simply dispersed into the air, swept away by the winds, and, because of its relatively small amount compared to the volume of the atmosphere, it had no noticeable effect on the environment. Wastes dumped into the rivers were washed away quickly and diluted to the point where no effect on water purity could have been observed, had anyone been able to measure it. Wastes left on land decayed by spontaneous chemical process: the balance of nature was preserved.

The situation today is not so simple. The problem is due to both quantitative and qualitative changes in the wastes we are producing. The natural cleaning processes are slow and can take care only of certain amounts and certain kinds of wastes. Today's obnoxious fumes produced by civilized society are too much for the atmosphere to disperse, especially in heavily settled areas. The rivers are not able to dilute and degrade sufficiently the variety of effluents dumped into them, with the deterioration of water quality visible to all. And one would have to wait a long time for the heaps of junked automobiles quietly to rust away and blend in with the earth's crust again as ores. This just will not happen because we are throwing them away faster than nature can degrade and absorb them. Further, plastics, aluminum, and certain other modern materials do not rust away as easily as iron did. Another example of the qualitative change in our wastes are the nonbiodegradable detergents. All these are well known to everyone interested in ecological problems.

We have to stop emitting waste materials into the air and waters, and stop piling up solid junk. This is a twofold problem. In-so-far as gaseous emissions to the atmosphere (other than water) exist in an electrochemical economy, they must be purified to the point where natural processes are able to take care of the remainder. If there are any effluents still dumped into rivers and oceans after electrochemical recycling processes have been introduced, they have to be pure enough that the water's natural degrading process can keep the desired water quality. This is the *waste treatment problem*. There is, however, also another problem: what to do with the materials removed from the effluents and with solid junk. This is the *waste disposal problem*.

In this chapter we present a few electrochemical processes which can be applied to the waste treatment problem. The waste disposal will be treated separately, in the following chapter.

7. 2. Electrochemical Treatment of Municipal Sewage

Municipal sewage is one of the largest single items as liquid effluent. Not too long ago, these wastes (mostly household sewage) were dumped into our waterways without treatment. Later some treatments ("primary treatments") were developed, usually consisting of a simple filtration or sedimentation to remove the solids, but most of the sewage was still ejected directly into the nearest river. Today, most sewage plants have also a secondary treatment, which consists of an oxidation step of the wastes. Plain aeration or contact aeration was the first method. Air was simply pumped into the treatment tank, sometimes filled with crushed stone to increase the area of contact between the air and the sewage. These methods required huge tanks (long aeration times) and still produced fairly poor treatment (extent oxidation of organic material, about 75%). More recently, the often used technique is the so-called activated sludge method. The sewage is aerated by mechanical agitation, or by the use of compressed air in the presence of a sludge containing bacteria which oxidize the organic materials in the sewage; the treated sewage is settled; the liquid is discharged while a part of the sludge is returned to the treatment tank. This method requires less land area than the previous ones and results in a 90–95% oxidation of the organic matter.

This brief survey of nonelectrochemical methods of sewage treatment was given for two reasons. The first was to show that they are all basically oxidation processes, something which, by its nature, can be done electrochemically (*cf.* Section 4.3 and Appendix II). Secondly, it should be noted that none of the treatments removed all waste constituents or oxidized all organics, and still they sufficed for some time to keep our waters clean. The rivers have a natural way of destroying sewage in addition to the dilution effect (basically a bacterial oxidation process similar to the activated sludge treatment, except, of course, in a much more dilute form) and, as a matter of fact, sewage can be an important food supply for aquatic life. It would be therefore unrealistic (and wasteful) to purify our sewage to the ultimate and dump triple-distilled water into our rivers. However, since the self-purification capacity of rivers is finite (and also variable from location to location), the larger the volume of the sewage discharged, the better the treatment needed.

In the short description of electrode reactions in Section 4.3, it was emphasized that at one of the electrodes the reaction is always oxidation (see also Appendix II), the type of reaction we need in sewage treatment. Indeed, electrolytic sewage treatment plants already were used at many locations in America (e.g., Santa Monica, California; Oklahoma City, Oklahoma; Elmhurst, New York; Allentown and Easton, Pennsylvania). In these plants, the sewage was first treated with lime, followed by an electrolysis with steel electrodes with a residence time of the order of a minute (compared to many hours of aeration time required for the nonelectrochemical methods), then settled and the clear liquid and sludge separately discarded. The plants operated well, their advantages being the small land area requirement and odorless operation. The treatment costs, however, were higher than those of the older methods and hence they were not developed. With increasing land values, due to population growth, and the increasing availability of low-cost electricity in an all-electric economy, the cost comparison could turn in favor of electrolytic treatment.

Another sewage treatment method is chlorination (chlorine gas is a strong oxidizing agent). This is applied as a pretreatment to decrease odors, or as the main treatment step, or as a final treatment (after aeration) to sterilize the treated sewage before discharging into the river. The chlorine gas, as will be discussed in Chapter 9, is made by electrolysis of salt (sodium chloride) solutions. Chlorination, therefore, can be considered as an indirect electrochemical method. There are more direct approaches. If chlorine can be made at the location, the process can be made more economical by eliminating the hauling costs, especially if cheap salt (e.g., the ocean) is near by. Such plants have been tried, e.g., in Norway, in Milan (Italy), and on Guernsey Island in the English Channel. The latter has been in operation for several years, providing satisfactory service. There are two ways to accomplish the treatment: the sewage can be mixed with seawater and electrolyzed, or the seawater can be electrolyzed in itself and the bleach solution* mixed with the sewage in treatment tanks. The reactions taking place in the electrolytic sewage treatment are complex. The organic matter can be oxidized directly at the electrode, the water will be electrolyzed with the produc-

---

*On electrolyzing salt water, the following overall reaction takes place:

$$2NaCl + 2H_2O \rightarrow Cl_2 + 2NaOH + H_2$$

if the chlorine and the caustic are allowed to mix, bleach will form:

$$Cl_2 + 2NaOH \rightarrow NaOCl + NaCl + H_2O$$

tion of oxygen (*cf.* Section 6.3), which in turn will oxidize the sewage; and in the case of seawater electrolysis, chlorine, caustic, and bleach are formed, which, again, will react with components of the sewage. In all probability a mixture of these appears, depending on local conditions.

In the United States, chlorine or hypochlorite is used as a disinfectant for the sewage plant effluent (after the normal treatments of sewage), especially if the plant discharge is close to beach areas. The oxidizing material is hauled to the treatment plant. Recently, several manufacturers made test installations of on-site hypochlorite generation cells, which would not only reduce costs (cutting it by as much as half), but eliminate the dangers in the hauling of liquid chlorine or concentrated hypochlorite solutions through heavily populated areas. The flowsheet of such a system is shown in Fig. 7.1.

Other disinfectants have been suggested: hydrogen peroxide and ozone. Hydrogen peroxide can be produced by electrolysis, through the oxidation of sulfuric acid to persulfuric acid and decomposition of the latter with water.* Ozone production is carried out from oxygen in a high-voltage discharge tube (a gaseous electrochemical process), and on-site generators for sewage treatment are being tested. Disinfection of water (not sewage) has also been suggested by the use of dissolving silver anodes, the resulting Ag^+ ion being poisonous to most microorganisms in minute concentrations.

Other possibilities exist for electrochemical engineering advances during the coming decades. Why have pipes to take sewage from houses—can it not be locally processed electrochemically? More than 60% of solid human wastes are cellulosic. Cellulose can be oxidized to CO_2 electrochemically. It seems likely that electrochemical conditions can be found for oxidizing the rest to O_2 and CO_2. What would the counterelectrode reaction be? Would it be acceptable to run the cell with H_2 evolution at the

---

*The reactions are as follows:

$$2HSO_4^- \rightarrow H_2S_2O_8 + 2e$$
$$H_2S_2O_8 + 2H_2O \rightarrow 2H_2SO_4 + H_2O_2$$

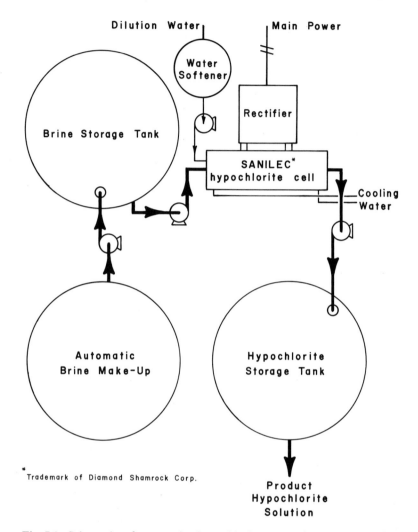

Fig. 7.1. Schematic of an on-site hypochlorite generating system at the Mentor–Willoughby (Ohio) Wastewater Treatment Plant (Courtesy Diamond Shamrock Corp.).

other electrode, bleeding the H_2 into the general supply to the house? Or might an air electrode deal with the electrochemical oxidation of sewage as a fuel cell process?

A much more far-out possibility may be mentioned as research-worthy. In the above suggestion, the eventual product of the sewage consumption would be CO_2. Its rejection into the atmosphere as a consequence of *sewage* treatment is not ecologically objectionable because it is the return path in the cycle of photosynthesis (the CO_2 will be taken once more to build plants which will feed man, who will form CO_2, etc.). However, it is interesting to note that one can reduce CO_2 to formaldehyde and from this—with atmospheric N_2 and an enzyme—form protein. The cycle could be completed locally in the home.

Lastly, a further stage in electrochemical oxidation would produce methanol, CH_3OH, and this is one of the fuels we could use in the interim, before we get abundant H_2, for running cars. Of course, if the fuel for cars driven by fuel cells were *derived* from CO_2, there would be no ecological objection to using it in a fuel cell and rejecting CO_2 back into the atmosphere again.

These aspects of possible sewage treatment are for future research. They stress ecologically helpful electrochemical possibilities.

7.3. Destruction of Industrial Wastes by Electrolysis

The oxidizing power of electrolysis can be used not only for the treatment of municipal wastes, but also in case of industrial effluents. As an example, the treatment of cyanide-containing waste streams will be discussed. Different cyanide compounds are commonly used in the metal finishing industry, and the wastes have to be completely freed of cyanide before discharge (only 0.2 part per million or less is allowed) because of their toxic nature. One way to accomplish this is the electrolytic oxidation of cyanide, the products of which are harmless nitrogen and carbon dioxide* (the amounts of carbon dioxide so released into the atmosphere are minute compared to the amount produced of fossil fuels and cannot be considered ecologically dangerous).

---

*The overall oxidation on the electrode is

$$2CN^- + 8OH^- \rightarrow 2CO_2 + N_2 + 4H_2O + 10e$$

A difficulty with a scheme as that described above is that if the pollutant has to be nearly completely removed, its concentration will be low toward the end of the process, thereby slowing down the rate of the reaction. (Remember, in an electrolytic oxidation, the material to be reacted has to get to the electrode surface to give up its electron and the velocity of this transport process to the electrode is proportional to the concentration of the species in the solution. These transport aspects of electrode reactions will be treated in Appendix III). A way out of this problem is to generate an oxidizing species electrolytically in excess and let this oxidizing agent mix with the cyanide solution in a holding tank where a chemical reaction can take place, destroying the cyanide. An oxidizing agent useful for this purpose is bleach (hypochlorite),* the electrolytic production of which from salt solutions has already been mentioned in Section 7.2 in connection with sewage treatment. The hypochlorite could be generated electrolytically and added to the waste stream, or the cyanide-containing waste could be directly electrolyzed after some addition of salt. In this latter case, not only the cyanide, but also the metal ion constituents of the wastes can be removed in one step; then it is mostly chlorine which is oxidizing the cyanide, and the caustic, instead of forming hypochlorite, precipitates the metals as insoluble compounds, or the metals can be directly plated out (*cf.* Section 7.4). Of course, chlorine gas or hypochlorite could again be bought for this treatment instead of producing it *in situ,* but at a higher cost, especially for the small plating shop operation.

7.4. Metal Ion Removal by Electrolysis

Up to now, we have discussed only the oxidation processes taking place during electrolysis. As we have shown in Section

---

* The reaction in this case is

$$2CN^- + 5OCl^- + H_2O \rightarrow 2CO_2 + N_2 + 5Cl^- + 2OH^-$$

Compare this reaction with that of the anodic oxidation; here it is the oxidizing agent (OCl^-) which picks up the electrons directly, rather than the electrode.

4.3 during electrolysis both oxidation and reduction take place, but spatially separated from each other; therefore, the oxidation process on one of the electrodes is always accompanied by a reduction process at the other electrode. A main application of this phenomenon is the reduction of metal from solutions in the form of solid metal, as exemplified by the removal of iron from the steel industry's pickle liquor:

$$Fe^{+++} + 3e \rightarrow Fe$$

(ion in solution) + (electrons in the electrode) → (solid metal) Such a process can be used for most metals. Note here that the metal is regained in pure form; furthermore, it is possible to separate individual pure metals from a solution containing a mixture of them. But this takes us into the area of material recycling, which will be discussed in the following chapter.

Since during the electrolysis both oxidation and reduction take place simultaneously, the idea is natural to try to use them concurrently for the removal of some oxidizable and reducible constituents from the waste stream at a considerable savings in cost as compared with that of two separate processes. There are difficulties inherent in such a scheme. The conditions favoring the oxidizing of certain wastes (e.g., temperature, concentration, flow rate, acidity, etc.) may not coincide with the optimum conditions for the reducing process (e.g., removal of a metal ion). Furthermore, the amounts of the oxidable and reducible waste components in the effluent may not be always equal. It is, therefore, quite probable that in a concurrent removal, neither of the processes could be operated at maximum efficiency, but the savings (both in capital and operating expenditures) in having only one single process may be large enough to overcome the loss of efficiency.

7.5. Waste Treatment by Electrodialysis

The field of electrochemistry involves not only the electrode reactions (utilized in the previously discussed oxidation and reduction treatments), it is also concerned with the movement and transport of electrically charged species (ions) in solutions. This area of electrochemistry (solution electrochemistry, or ionics) is also applicable to pollution abatement.

Many chemicals will, upon dissolution in water, dissociate into charged particles, causing the solution to be electrically conductive, e.g.,

$$NaCl \rightarrow Na^+ + Cl^-$$

When a direct current is passing through the solution, the positive ions (cations) will move toward the negative electrode and the negative ions (anions) toward the positive electrode. Consider now a cell which is built of numerous compartments, each separated from the other by a special plastic membrane, a so-called ion exchange membrane, which has the characteristic of allowing the passage only one kind of ion (either cation or anion). The basis of operation of such membranes is outside the scope of this book; let it suffice that they are a practical reality, commercially available at reasonable prices. Such a cell is shown schematically in Fig. 7.2. Let us examine what happens in compartments 2 and 3 during the passage of an electric current, supposing that the compartments were originally filled with the same salt solution. As the current passes, the anions from compartment 2 will move toward the left into compartment 1, and cations will move toward the right into compartment 3. The membranes separating the compartments are of such character as to allow the passage of these ions. What about compartment 3? The positive ions would like to move, under the influence of the electric field, to the right into compartment 4, but they cannot since the anion exchange membrane separating the compartment will not allow the passage of the positively charged ions. The anions of compartment 3 would prefer to move toward the left but are again held back by a membrane which passes only cations. What is the result? The solution in compartment 2 will become more and more dilute in salt as time passes, while that in compartment 3 will become more and more concentrated. The same thing happens in every neighboring compartment pair throughout the stack (in the end compartment there is also electrolysis occurring at the electrodes). The solutions are, of course, not stagnant as suggest-

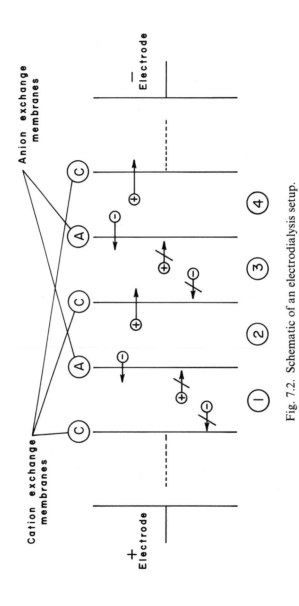

Fig. 7.2. Schematic of an electrodialysis setup.

ed by Fig. 7.2, but are flowing through the compartments at any desired rate.

The application is obvious. From a large volume of a polluted water stream, containing a given salt in relatively low concentration (but not low enough to be discarded without treatment), the pollutant can be removed and concentrated into a much smaller volume of highly concentrated solution, while the original polluted stream, now containing much less salt, can be discharged as treated waste. What to do with the concentrated solution is another problem, but not unique for this technique. In whatever way we remove pollutants from an effluent, the removed waste will have to be taken care of somehow: this is the problem of waste disposal. A specific application of the above process, which has been developed during recent years, is the treatment of paper mill wastes.

7.6. Waste Removal by Electroflotation and Electroflocculation

It has been mentioned before that one type of electrochemical technique is to generate a reagent at the electrode which can further react with the waste material. Such an electrochemically produced "reagent" can, however, be used also in other ways not only to perform oxidation or reduction.

In case the water is polluted by small solid particles or by a liquid immiscible in water, a purification technique called flotation can be used which is the inverse of the familiar sedimentation process. The foreign particles can be removed by forcing them, with a stream of fine gas bubbles, to rise to the surface, from where they can be skimmed off. The needed gas stream could be produced by electrolysis of the water. Typical wastes which are treated with this method are oil in water immersions (oil industry, food and dairy industry) and fibrous materials (asbestos, textile, glasswool, etc.). The action of the electrolysis has not been fully investigated yet; in addition to producing the stream of small gas bubbles needed in the process, it can have other effects. A suspension of this kind is often stabilized by the particles being electrically charged (*cf.* Appendix I) and electrolysis could neutralize some of these charges, aiding the coagulation and separation from water.

Another way of using the electrochemically generated reagents is as coagulants or flocculants. The small suspended solid particles are very slow in settling and large tanks would be required. A standard treatment is to add reagents like aluminum sulfate or ferric chloride, which act in two ways: the large positive charge of the ions (Al^{+++}, Fe^{+++}) can neutralize the negative charge of the particles (*cf.* Appendix I), helping their coagulation (otherwise the similarly charged particles repel each other) and therefore sedimentation; also, these metals form hydroxide precipitates which are fluffy, with a high surface area, by nature, and this so-called "floc" entraps and sediments the fine suspended solids. Now aluminum or iron can be added to the solution by electrolytically dissolving the metal electrodes (the reverse reaction of the electroplating discussed in Section 7.4). This is an expensive way of adding coagulants to the waste and probably will not be economical to use in itself. If the proper electrodes are used however, it can be a useful side effect in the electrolytic treatment of sewage or in electroflotation. In one proposed electrolytic sewage treatment, aluminum is electrolytically dissolved to precipitate the phosphate present from detergents.

7.7. Electrofiltration

Many different waste treatment techniques involve a filtration step somewhere in the process, either for removal of solid wastes or in cases where some components are removed by chemical precipitation. Filtrations are normally troublesome; the solid material to be removed will sooner or later plug up the filter, necessitating periodic cleaning. A new technique, electrofiltration, has recently been suggested, which would ameliorate the situation. It is based on a fact, mentioned before, that the small particles to be removed have a net electric charge (see Appendix I). If two electrodes are placed in the filter and an

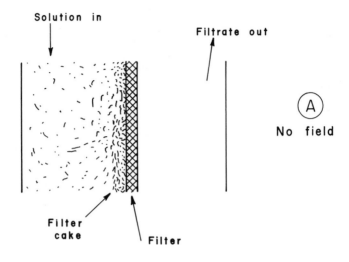

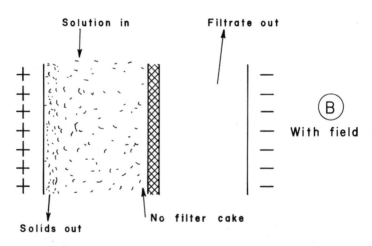

Fig. 7.3. Comparison of electrofiltration with normal filtration.

electric field applied in the appropriate direction, the particles will be driven away from the surface of the filter avoiding its plugging. The idea is illustrated in Fig. 7.3.

7.8. Electrostatic Precipitators

This method is not strictly electrochemical, in the classical sense, since it involves gas-phase operations at many thousands of

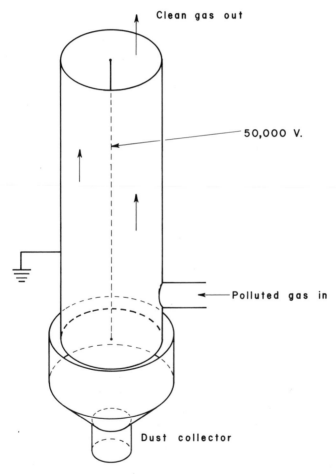

Fig. 7.4. Schematic of the electrostatic precipitator.

volts; a brief description is included here for the sake of completeness. It is a standard method for the removal of dispersed solid particles or liquid droplets from gases; it is used for purification of industrial gases from dust, flyash, mists, and smoke. The principle of its operation, and also the equipment itself, is simple; it is illustrated schematically in Fig. 7.4. The polluted gas stream flows through a grounded pipe containing in its axis a high-voltage wire electrode so as to create a field strength of several thousand volts per centimeter. The particles become electrically charged in this high field and migrate to the wall, where they discharge and collect; the pipe is periodically mechanically rapped to dislodge the collected rubbish, which then falls into the dust bin. This method of gas purification can be very effective (99 % or higher). However, the cost is in the vicinity of $1 million per chimney.

7.9. Closing Remarks

A number of different electrochemical techniques have been described in this chapter which can successfully be used in pollution abatement, in the cleaning up of our liquid and gaseous effluents to preserve the quality of our waterways and atmosphere. There is one feature common to all of them (as a matter of fact, common also to nonelectrochemical waste treatments), that is that they need extra power, not previously needed for the processes in operation. If the exhaust of intenal combustion engines does get partly cleaned up, the mileage per gallon of gas will go down. When the fume emission of a power plant is reduced, some part of the power produced will be used up in the effluent purification steps. Treatment of municipal and industrial waste (either electrochemical or by other methods) will all require some form of power input. When new energy sources make electricity abundant, this will be a small price to pay for the clean environment. The situation in the near future, however, is going to be increasingly difficult, particularly in the period between *ca.* 1980 and the time at which the new power sources are ready. It is a fact sometimes overlooked by ecologically oriented people that from the time at which the realization for a new process comes into play until its completion may be typically 15–20 years.

One of the ways in which this situation could be ameliorated is to attempt to create fuel cell processes out of as many electrolytic clean-up processes as possible. For example, it has been shown that SO_2 (a frequent pollutant in chimney smoke) can be oxidized to sulfuric acid by a fuel cell process: the by-product is electricity (see Section 8.5).

CHAPTER 8

Electrochemical Methods of Waste Disposal

8.1. Waste Disposal and Recycling

The problem of waste *disposal* was separated in the previous chapter from that of *effluent treatment*. This problem has been around throughout the ages, and it was formerly simply solved, mainly by letting nature take care of it. Unwanted materials were left to ferment away, i.e., to become largely CO_2 by the action of bacteria. While the total amount of waste was small, this was acceptable, and the ecological situation was not disturbed. Often, burning was used to help the decay process, transferring most of the wastes into the atmosphere. In an ecologically conscious society, one which has to deal with and reverse disturbances of nature's balances, these methods are not acceptable. While natural decay still is a way of waste disposal, it must be limited in extent, and handled in a more careful way. The wastes are now compacted, deodorized, used in sanitary landfills, etc.; efforts are made to avoid odors and atmospheric pollution (burning is unacceptable for the future, even with purification of the effluent gas, because of increasing still further the unbalanced carbon dioxide emission). Because of the extra efforts in handling, and because of the increasing

117

volume of waste per person generated by modern societies, costs spent in the treatment of wastes are increasing. The cost increase is leading to a situation where reuse of the waste, often after some suitable chemical transformation, begins to look feasible, especially if one includes (as one now must try to) the cost of the ecological and esthetical effects, however difficult it is to put a finger on a dollar value for these. This leads us to the ultimate method of waste disposal: *recycling,* which may be increasingly used in the future.

There is another important reason for man to move toward a total recycle economy: our consumption of raw materials. In the past, our mineral resources were handled as though they were inexhaustible. To be sure, they were not destroyed in the physical sense, but they were made unreusable in any practical sense. The burning of coal results in carbon dioxide in the atmosphere transferring very slowly into the oceans, eventually to form carbonate deposits; the carbon is still there but hardly usable as a fuel any more. When an iron article is left to rust, it returns to the form it had before extraction, iron oxide; but while the iron ore was in a concentrated deposit, the result will be dispersed on the earth's crust at such low concentrations as to make its mining uneconomical. Our resources are becoming increasingly dispersed on the earth, and sometimes chemically converted, resulting in "deposits" not economically recoverable, and so, in effect, lost. When mankind's total use of minerals was proceeding at a slow rate, they must have seemed inexhaustible, but today the utilization of our raw materials is skyrocketing, just as the use of energy discussed in Chapter 2. If the growth pattern of the last few decades continues unchanged, we shall exhaust many of our metal ore resources (e.g., copper, aluminum) in less than half a century, and even the more abundant metals (like iron) will not last more than a couple of centuries.

These two trends, the rising cost of waste disposal and the exhaustion of economically minable raw materials, will force us more and more to consider our waste piles as mines and operate a closed-cycle economy. We already have some limited experience in closed-system living from space flights, where, e.g., the small oxygen and water supply has to be regenerated for reuse; such procedures will be increasingly common as the mission time increases (e.g., Skylab) and recycling becomes accepted as a normal procedure. But, in effect, we are all living in a closed spaceship, the Earth, which has no material input (except about one hundred tons per day of meteorites), and we shall have to make do on fixed resources, the steady state of the recycling quantities. This has always been the situation, but it never seemed important to us until recent years of this century because of the slow rate of use of our reserves; we never saw the bottom of the barrel, nor knew where it would be found. In the future, however, we cannot afford the luxury of over-looking the fact that we can no longer just go on and on, but have to put back and recover everything we use.*

Material recycling is an area of technology not much developed as yet. There are, however, some exceptions: more than half of all lead used is provided by recycling, also considerable amounts of copper and iron are presently reused, but we are still far from total or nearly total recycling. (A 100% recycling will be difficult and some make-up will be needed from "primary" sources for a long time; this is a reason to convert to recycle systems as soon as possible, before greater exhaustion of our primary sources.) This recycle technology will have to be developed in the next few decades and we shall show in the following sections that electrochemical techniques are available (some of them uniquely qualified) to tackle this job.

8.2. Recycling of Metals

The recycling of metals is an area where electrochemistry is uniquely applicable and related technology is already developed. Many metals are presently produced, purified (refined), or applied as coatings using electrochemical techniques, as will be discussed in more detail in Chapter 9. Their use should

---

* One could say we have been spending our capital as though we were such multimillionaires that reckless spending could never make us poor. But from now on we shall have to live on the income on what is left of the capital.

increase in the future, partially because the electrochemical approach is less polluting than the corresponding chemical one. The removal of metallic constituents from waste streams has already been mentioned in Section 7.4. The metal is not only removed from the waste in that process, but is reclaimed in a pure metal form. It is interesting to add here that the metal can be regained, if so desired, in special forms imparting extra value to the product: fine metal powder (with controlled particle size), thin foil, or wire are commercial electrolytic products. The market for these special forms is at present limited, and most metal probably will be reclaimed as heavy sheets or ingots from electrodes. This technique is applicable to solutions of a metal which is predominantly present in the solution, such as plating-shop waste streams, spent acid used to pickle metallic articles, etc. If the solution is too dilute, an electrodialytic concentration step (see Section 7.5) may precede the electrolysis.

The bulk of the metal wastes to be recycled, however, is not in solution but in solid junk form. Furthermore, it seldom consists of one metal only, for few appliances are made of, for example, iron only or aluminum only. They usually consist of several different metals joined together. The metallic content of a typical car, for example, is shown in Table 8.1. This is the kind of mixed metal scrap facing the technologist embarking upon recycling. This is not a special case; many metals are complex alloys, each containing many constituents. Electrochemistry offers a way for the separate reclaiming of each metal.

Table 8.1. Metallic Content of a Typical Car

Metal	Pounds
Iron	3000
Copper	32
Zinc	54
Aluminum	51
Lead	20
Nickel	5
Chromium	5

It is known that some metals can be oxidized more easily than others. The noble metals, like gold, will not oxidize in air; others, like iron, will slowly rust; while metals at the other extreme, like sodium, cannot be kept in air for any length of time without oxidation. These differences toward oxidation–reduction processes exist when the process is carried out electrochemically and are the differences which can be utilized for the separation of the metals. The oxidizing or reducing power of an electrode can be controlled by changing the potential* which it is made to take up by the adjustment of an outside source of power, and therefore, a separation of metals can be achieved. For example, a machine part containing both iron and copper pieces can be dissolved electrolytically (oxidized to ions in solution) at a certain potential so that only the iron† will dissolve and the potential will have to be increased to dissolve the copper. Conversely, from a solution containing both copper and iron, in dissolved form, the copper can be plated out (reduced from the ionic to the metallic state) first by itself,‡ and the iron will plate out only after a change in the potential; if the potential change is made only after all the copper is removed from the solution, the two metals can be plated out separately, preferably onto separate electrodes, achieving a separation. In principle, therefore, solid metallic junk can be separated into its pure components by either selective dissolution or selective plating. Selective dissolution encounters a number of difficulties in potential control; erroneous potentials can occur because of loose connection of different parts in the machine or different distances of the parts from the control electrode. (This aspect, the potential drop in the solution, will be elaborated further in Appendix III.) Selective *plating* out from the solution is the more practical possibility.

---

* This is the potential difference between the electrode and the solution surrounding it, and *not* the cell voltage as measured between the two electrodes of the electrolyzer. This will be treated in more detail in Appendix II.

† The iron, being the less noble metal, is more easily oxidized.

‡ The copper, being the more noble metal, is more easily reduced (less easily oxidized) than iron (oxidation and reduction are complementary processes).

A reclaiming plant may be envisioned as follows. The metallic junk is shredded into small pieces, the paint burned off (for otherwise there is no ionic connection to the solution), and sections compacted into flat electrodes; it is then nonselectively dissolved into acid electrolyte from which it is selectively plated out at a succession of potentials each chosen to obtain a specific metal on a specific substrate. Could new car bodies perhaps be directly formed in the bath? Alternatively, the metal could be dissolved in acid and the metals selectively plated out from the solution while the acid is regenerated (see Section 8.5), or the evolved H_2 used to produce electricity. Preliminary cost estimates have shown that metals from junked cars could be reclaimed, with this type of a process, at a profit; esthetic gains, for example, the elimination of piles of junked cars accumulating around our cities, are not included in the calculation.

8.3. Water Recycling

Waste water treatment has been discussed in Chapter 7. It was emphasized that treatment should produce such an effluent quality that the water could be run into our waterways without danger to aquatic life. Further, natural purification processes occur in the water itself; then a municipal water works will draw water from the river and after further purification will distribute it for general use. This, in itself, is a recycling process; the water of a river can be used several times over before it reaches the ocean. The loop could be completely closed and the two processes integrated into one operation, that is, that the effluent of the sewage plant could be directly processed further for reuse. This, of course, would make the requirements of sewage treatment more stringent as far as decomposition of organic wastes is concerned and would add an extra requirement of removal of salts added during treatment (e.g., bleaching). Electrodialysis would be a viable step for the removal of dissolved salts (Section 7.5); the other steps of the treatment (oxidation, sterilization, etc.) could remain as discussed before, except with increased efficiency.

Electrochemical desalination of sea water, and particularly brackish water, may be mentioned here: it can be considered a recycling process since most of our "used" water ends up in the oceans. Plants based on the electrodialysis process (Section 7.5) are a contender in this field. Such plants are presently under large-scale trial, in comparison with other methods. The cost of electricity is an important factor here, and this will drop sharply as the size of the atomic reactors used to produce it increases. One such desalination plant is sketched in Fig. 8.1; this produces 1.6 million gallons of fresh water per day, from brackish water (reducing the salt content from 1.3 grams per liter to that fit for drinking) at a total (operating + capital) cost of 29 ¢/thousand gallons. The production of drinking water, in the hydrogen economy, as a by-product of power generation (see Section 6.5) is a kind of more remote or indirect desalination process.

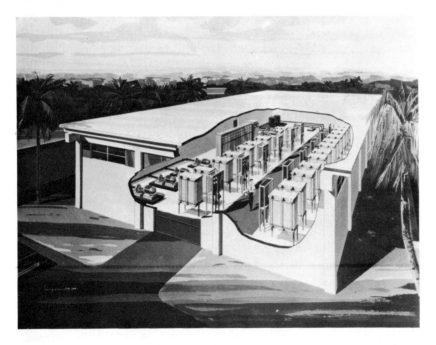

Fig. 8.1. Artist's drawing of a desalination plant, located at Siesta Key, Fla., using the electrodialysis process. (Courtesy Ionics Inc.)

8.4. Air Regeneration

At present, this is a special problem relating to closed environments such as spaceships, submarines, etc. It may, however, have important implications for the future, when it can be visualized that whole towns will have integrated climate control systems (cooling or heating, as required), perhaps under a plastic dome. In this case, it may be more economical to regenerate oxygen from carbon dioxide, rather than to draw in fresh air. A number of electrochemical systems have been worked out for this purpose.

One would use a molten-salt (lithium carbonate) electrochemical cell: the carbon dioxide is dissolved in the molten electrolyte and decomposed during the electrolysis to yield oxygen gas and solid carbon. Another process would electrolyze water to produce oxygen and use the hydrogen by-product to reduce the dioxide to carbon and water in a catalytic process. In other schemes, the carbon dioxide would be reduced not to solid carbon, but to some hydrocarbon product (e.g., methanol) which can be used as a chemical raw material (see Section 7.2), while oxygen is produced at the anode. Other processes have also been suggested, but these suffice to show the applicability of electrochemical techniques to the generation of fresh air. They provide a process which, in the future, may become a standard municipal utility, just as water and sewage plants are today.

8.5. Miscellaneous Electrochemical Recycling Processes

Sulfur dioxide is one of the more serious pollutants of the atmosphere, originating mainly from power plants burning sulfur-containing fuels. As a pollutant, it is a health hazard: high levels of SO_2 occur during thermal inversions and produce aggravation of symptoms for people suffering from asthma and chronic respiratory and heart diseases, and the mortality rate from these diseases has been correlated with the sulfur dioxide content of the atmosphere. Also, it may cause damage to vegetation and property (corrosion of metals, marble, and other stone structures). The removal of sulfur dioxide from flue gases (or the removal of sulfur from the fuel before burning) will be a heavy cost burden on power plants. If the sulfur could be regained in some usable form (e.g., sulfuric acid), part of the extra cost could be avoided. It has been estimated that, for a highly industrialized country, the conversion of all the sulfur emitted from power plants (assuming 1% sulfur in the fuel) would produce about 30% of the sulfuric acid need of the country. Such a process would not only purify the atmosphere, but would also give a useful and needed chemical product.

Several electrochemical schemes have been proposed, and some tested on a pilot-plant scale, to convert the sulfur dioxide from flue gases to sulfuric acid. The process operates by using a scrubber solution to absorb the SO_2 from the gases (recently the use of molten-salt scrubbers has been suggested; this would allow operation at high flue gas temperatures, which is needed to produce a good chimney effect), with the solution then treated electrochemically, to form sulfuric acid and to regenerate the absorbing solution. The reactions and the processes are complex (usually requiring multicompartment cells with ion exchange membrane separators) and will not be discussed here. The process would use electrical energy, but it would be possible to operate in such a way as to use low-cost (off-peak hour) electricity, by storing the absorbing solution during the day and regenerating it at night.

A different process would utilize the sulfur dioxide as a fuel in fuel cells and would produce not only sulfuric acid, but also electricity as a by-product. The overall cell reaction is

$$O_2 + 2H_2SO_3 \rightarrow 2H_2SO_4*$$

* Sulfur dioxide would be oxidized on one electrode:

$$SO_2 + H_2O \rightarrow H_2SO_3$$

$$2H_2SO_3 + 2H_2O \rightarrow 2H_2SO_4 + 4H^+ + 4e$$

this and the concurrent reduction of oxygen (from air)

$$O_2 + 4H^+ + 4e \rightarrow 2H_2O$$

result in the above overall reaction.

The theoretical voltage of the cell is a little over 1 volt, but because of polarization lossess (see Appendix II) probably no more than half of this could be utilized.

Another recycle process which may be mentioned as an example is the regeneration of chrome plating baths. Chromium plating is extensively used as a decorative and protective coating on metal parts of automobiles, appliances, etc., and chrome chemicals are widely used oxidizing agents (e.g., tanning of

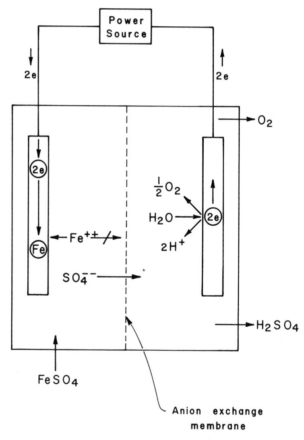

Fig. 8.2. Schematic of pickle acid regeneration. Overall reaction:

$$2FeSO_4 + 2H_2O \rightarrow 2Fe + 2H_2SO_4 + O_2$$

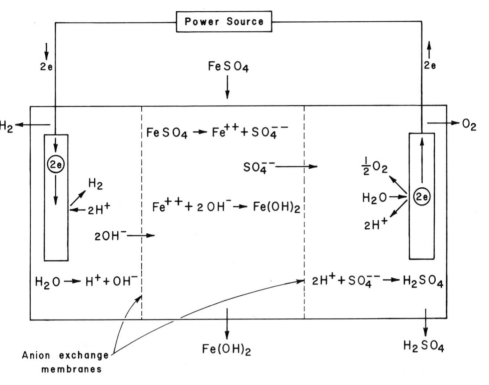

Fig. 8.3. Schematic of pickle acid regeneration. Overall reaction:
$$2FeSO_4 + 6H_2O \rightarrow 2Fe(OH)_2 + 2H_2SO_4 + 2H_2 + O_2$$

leather). The spent baths, still containing large amounts of chromium, are a difficult waste to dispose because of the poisonous nature of the chrome salts. Instead of destroying the waste, they can be electrochemically regenerated; the process not only solves the waste disposal problem, but regains most of the chrome.

As a final example, the regeneration of pickle acids used in steel mills to remove scales can be mentioned. The "spent" acid contains large amounts of iron and considerable free acid; it is a difficult waste to handle. Several suggestions have been made for reclaiming the acid electrochemically. Some would also recover the iron (Fig. 8.2); some would precipitate it as a hydroxide (in this case, the disposal problem of a large volume of acidic solution would be reduced to disposal of a much smaller volume of noncorrosive solids); some would reclaim the acid as such; and some (if hydrochloric acid is used in place of sulfuric acid) would produce chlorine gas as a useful by-product. Two schemes are shown in Figs. 8.2 and 8.3. The process shown in Fig. 8.2 uses a two-compartment cell separated by an anion exchange membrane which does not allow the passage of metal ions; this results in the reclaiming of the iron as metal and, since on the other electrode oxygen is produced from the water $(2H_2O \rightarrow O_2 + 4H^+ + 4e)$, the by-product hydrogen ions combine with the sulfate, giving sulfuric acid, which can be reused for pickling. The process shown in Fig. 8.3 uses a three-compartment cell with two anion exchange membranes. The acid is regenerated as in the first scheme, but the iron is discarded as hydroxide precipitate. The reactions are shown in the figure.

8.6. Conclusions

Recycling technology is, at present, the least developed among the waste treatment methods. Most of the processes discussed on the preceding pages are not in actual use, some are as yet only suggestions, although some have already been brought to pilot-plant level. Recycling must become the most important technique for waste treatment in the not too distant future— before 2000 A.D.—because of the threatening raw material shortage. The possible uses of electrochemical techniques is more widespread than was indicated by the few examples. All the techniques described for waste treatment in Chapter 7, and all the production techniques to be discussed in the next chapter, can be used in one form or another, and with increasing ecological consciousness, and decreasing electricity costs, their development in the near future seems reasonably certain.

It is noteworthy that the number of people the earth can support to 2100 A.D. at an affluent standard probably depends upon the steady-state "concentration" of raw materials. Until we calculate this in some detail, we shall not know what number of people we can eventually support.

CHAPTER 9

Electrochemical Manufacturing

9.1. The Role of Electrochemical Technology

The main message of this book is to show how electrochemistry can be utilized to help in solving problems connected with the disturbance of the balance of nature caused by present transportation and industrial methods. It is suggested that, through the availability of abundant, low-cost electricity, and the introduction of the hydrogen economy, electrochemical technology will become a major way for the production of goods; the recycling of metals; the handling of transportation; avoidance, measure, and cleaning up of pollution, etc. In the present chapter, one has to introduce electrochemical technology as it exists today; to show its applicability to a variety of products; and the possibilities for future developments, including the avoidance of pollutional problems through the use of electrochemical techniques. Few details will be described; the interested reader is referred to the books given in the reading list for more comprehensive descriptions and technical details.

9.2. Production and Purification of Metals

Electrowinning and purification of metals is a large chapter of electrochemical technology. The electrochemistry of these reactions was discussed in Sections 4.3 and 7.4. It was shown how metals can be produced in pure form through electrolysis of their salt solutions, and, it may be added here, from their molten salts. These reactions are useful not only in batteries and for removing metals from waste streams, but for the primary production of these metals: electrowinning. It was also shown in Section 8.2 that, by controlling the conditions of the electrolysis process, pure metals can be obtained from a solution containing a mixture of metals. This is especially favorable economically when small amounts of impurities are to be removed from a metal to produce a high-purity product, this is electrorefining, which is also a successful commercial operation; in these cases, a metal sheet electrode is dissolved into a solution from which the pure metal is then plated out, and both purification processes, controlled dissolution and controlled plating (cf. Section 8.2), are utilized simultaneously.

The list of metals produced or refined electrolytically is a long one, and includes most of the commonly used metals (except iron). A partial list is: aluminum, cadmium, chromium, cobalt, copper, gold, lead, magnesium, manganese, nickel, silver, sodium, tin, and zinc. To give some idea of the size of operations, some total production figures for the U.S.A. for 1971 are quoted: aluminum, close to 4 million tons; copper, 2 million tons; zinc, about 1 million tons; and magnesium, over 100,000 tons.

An example of the elimination of pollution through electrochemical technology can be pointed out here. Many metals (e.g., copper, zinc, lead) are found in nature as sulfides, and in the nonelectrochemical processes of metallurgy, the first step is invariably a smelting, where the sulfide is turned into oxide, and the sulfur is driven off as sulfur dioxide. The problems of sulfur dioxide emission into the atmosphere have been pointed out in Section 8.5. An acid leaching of the ore, followed by electrowinning of the metal, eliminates this atmospheric pollution, and the use of the electrochemical route has been gaining favor in recent years. The leaching of the ore itself is possible through an electrochemical operation. Pilot-plant studies are being conducted where the ore is leached by chlorine or hypochlorite produced *in situ* in a slurry of ore (cf. Section 7.2 for some of the reactions), and the resulting metal chloride solution can be used for electrowinning. In principle, the electroleaching and electrowinning operations can be combined, using the chlorine generated at one electrode to leach the ore and simultaneously depositing the metal at the other electrode. Application of these processes would eliminate the sulfur dioxide emission into the atmosphere.

One future application, presently under intense study, is the electromining of metals from the oceans. The "mining" and processing of manganese nodules from the ocean floors is in the pilot-plant stage today. One organization is planning processing facilities to turn one million tons of nodules per year into 260,000 tons of manganese, 12,600 tons of nickel, 10,000 tons of copper, and 2,400 tons of cobalt.

9.3. Inorganic Electrochemical Processes

Metals are not the only materials which can be produced electrochemically. A number of inorganic chemicals are manufactured by electrolytic processes. A partial list of these includes: chlorine gas, fluorine gas, sodium hydroxide, potassium hydroxide, sodium hypochlorite, sodium chlorate, sodium perchlorate, hydrogen peroxide and persalts, hydrogen and oxygen gases, and heavy water. The technology and electrochemistry of some of these processes is complex and will not be treated here; as an example, Fig. 9.1 shows a modern chlorine-caustic cell. The following figures indicate the 1971 U.S. production of some of these chemicals: chlorine, over 9 million tons: sodium hydroxide, over 10 million tons; sodium chlorate, nearly 200,000 tons; and fluorine gas, over 100,000 tons.

These chemicals may be unfamiliar to some readers since they are not everyday consumer products. They are, however, basic building blocks of many widely used materials such as plastics, sythetic fibers, paper, refrigerants, aerosols, dry-cleaning fluids, soaps and detergents, glass and ceramics, etc. Their widespread use is indicated by the above quoted production figures.

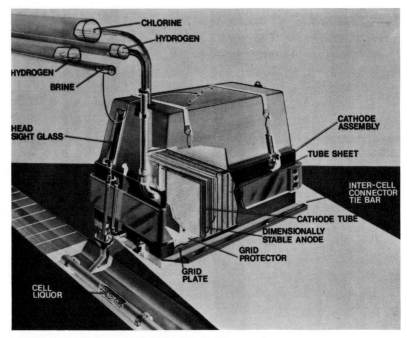

Fig. 9.1. A modern chlorine–caustic cell. The reactions are:

cathode: $2Cl^- \rightarrow Cl_2 + 2e$
anode: $2H_2O + 2e \rightarrow H_2 + 2OH^-$
overall: $2NaCl + 2H_2O \rightarrow Cl_2 + 2NaOH + H_2$

(Courtesy Diamond Shamrock Corporation)

9.4. Organic Electrochemical Processes

The list of organic materials produced electrolytically is not as impressive as those of metals and inorganics. An outstanding example is the production of adiponitrile, used for the manufacture of the widely used plastic, nylon (Fig. 9.2). Some other processes are the production of fluorocarbon compounds, tetraethyl lead, amino compounds, and sugar derivatives. The patent and chemical literature abounds in different electrolytic preparation techniques for a wide variety of organics, but only few were carried on to full-scale production. The reason for this underutilization is partly the price of electricity and partly the lack of fundamental understanding of the complex processes

Fig. 9.2. Cell room for the production of adiponitrile. The reaction at the cathode is

$$2[CH_2 = CH - CN] + 2H_2O + 2e \rightarrow [NC - (CH_2)_4 - CN] + 2OH^-$$

The reaction at the anode is

$$H_2O \rightarrow 2H^+ + \tfrac{1}{2}O_2 + 2e$$

The overall reaction is the dimerization of acrylonitrile to adiponitrile and oxygen gas by-product. (Photo courtesy Monsanto Company, and *Chemical Engineering Magazine*.)

taking place in the electroorganic reactions (so that, when something goes wrong, it is not so easy to cure as when the reaction is understood).

9.5. Electroplating and Electrocoating

An important part of electrochemical technology is the application of decorative and protective coatings on metal and, more recently, on plastic surfaces. The electrochemistry involved is the same as in metal winning: the parts to be plated are made the negative electrode in a solution of a metal salt, which is reduced electrolytically to form the metal coating. Practically all metals can and are used for electroplating, the most widely applied ones being copper, nickel, chromium, and zinc; they are used on many everyday items and there is no need to give here a list of applications. The coating usually serves a double purpose: it enhances the appearance of the parts and gives a corrosion protection for the underlying, less resistant metal (e.g., steel). It is probably less widely known that this metal plating technology has recently been extended also to plastic parts. In this case, first a thin conductive coating is applied by chemical means; once the surface is electrically conductive, the same plating technique is used as for metals. This application is mostly for decoration, to give a metallic luster to the plastic parts.

Another modern technique is electrocoating of metallic parts by nonmetallic coatings (paint). In this case, a water dispersion of resin particles is used, which was prepared in such a way as to leave the resin particles ionized (possessing an excess of electric charge) so that they will move under the influence of an electric field. At the surface of the metal parts to be coated, chemical changes (taking place due to the presence of an electric current) cause the resin particles to precipitate and form an adherent, uniform coating. The electrochemistry involved is complex and, at present, only partially understood. Depending on the type of coating used, the metal to be coated can be the positive or the negative electrode in the process. There are several advantages of this type of "painting" *vis-a-vis* the conventional processes. Some of these are: complex-shaped parts can be uniformly coated with paint, the current reaching into every crevice, hole, or corner; the water-based paints eliminate the inherent fire hazards and pollution problems of the organic solvents used in conventional paints; the process can easily be automated. A disadvantage is that only one thin coating can be applied by this technique, since an electrically conductive surface is needed. This is a new technology (about ten years old) which is now making fast inroads in the automotive and appliance industry. Presently, it is being used for the application of an undercoat of protective.paint before the application of the final coating. The use of this technology for the application of a one-coat final finish is being pursued.

9.6. Electrochemical Machining and Electroforming

In addition to production, refining, and coating of metals, electrochemical technology can be used to shape metallic parts by two different techniques, one depending on controlled dissolution, and the other on controlled deposition.

Electrochemical machining is a new technology (developed during the last 20 years) which uses selective electrochemical dissolution of the workpiece to form it to the desired shape. A direct-current electrical supply is connected to the "cell" consisting of the forming tool, the workpiece, and a conducting solution between the two (Fig. 9.3). The tool is moved forward at a speed of about a thousandth of an inch per second while the solution flows at a high speed in the thin gap (a high flow rate is needed to carry away all the dissolution products and the generated heat). The workpiece is dissolved electrolytically

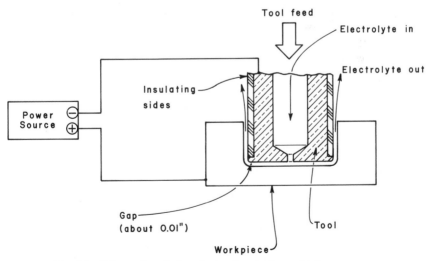

Fig. 9.3. Schematic of the electrochemical machining process.

according to the shape of the tool (note the insulation on the side of the tool in Fig. 9.3; it is there to avoid widening of the hole). This technique has developed together with the modern alloys, the shaping of which is costly, and sometimes impossible, with conventional machining techniques due to their extreme hardness. Further advantages of the electrochemical technique are that complex shapes can be produced in one operation, that the process can easily be automated, and finally that very thin sections can be produced with high precision, the machining of which would be impossible by conventional means.

Electroforming is the opposite process, based on electrolytic deposition of the metal. Deposition is carried out onto a form, the surface of which is specially treated to allow separation of the deposited metal once the desired thickness has been reached. The advantage of this method is that complex shapes can be produced with high precision, for which conventional production techniques would be costly. One of the largest applications, which indicates well the special characteristics of the technique, is the production of stampers for records. The record is originally cut, during the recording, into an aluminum disc from which a nickel "master" is made by electroforming. This "master" itself can be used to stamp out the plastic records, but one stamper is good only for about 2000 records. Therefore, it is used to produce a number of stampers through a succession of two more electroformings (the first one giving the negative of a stamper and the second the final stampers). To give the quality required in today's high-fidelity recordings, any imperfections in the groove must be less than one-thousandth of an inch, and this precision can be achieved after three successive copyings. Large numbers of other consumer and industrial items are produced by electroforming, such as electronic parts, electric razor heads, airplane engine parts, waveguides for microwaves, etc.

9.7. Corrosion Prevention

Corrosion, the destructive dissolution of metals, is with very few exceptions an electrochemical process and, as such, can be combatted with electrochemical means. How metals dissolve (oxidize) electrochemically was already discussed at some length in Section 4.3. For an electrochemical cell to be operative, one needs two electrodes immersed into a conductive solution, but cell action can also occur on one piece of metal covered by a thin film of moisture. Different parts of the metal piece can act as the two electrodes, and a moisture film is always present on the surfaces except in the driest of atmospheres (air pollutants, such as sulfur dioxide and carbon dioxide, dissolve in this moisture film and enhance corrosion because they increase the film conductivity and hence the flow of current between different parts of the metal).

Without going into much detail, it can be simply shown how electrochemical means can be used to stop this action. Consider only the metal dissolution reaction, resulting in the metal being dissolved as positive ions and leaving electrons in the metal. These electrons will then be consumed at the other "electrode" (some other part of the same metal) for a reduction process, for example, hydrogen evolution (Fig. 9.4). Since no gross deviations from electroneutrality can occur, all the electrons produced by the metal dissolution will have to be used up by the other electrode process. If, however, electrons for this reduction process are supplied by an outside source, the corrosion will stop since there is no "sink" for the electrons produced by the metal dissolution. This is an oversimplified picture, but gives the essence of one of the several electrochemical corrosion prevention techniques widely used in practice.

9.8. Concluding Comments

A cross section of electrochemical technology, not directly related to pollution abatement, was given in the preceding paragraphs to indicate its widespread applicability to production of industrial and consumer goods. It is emphasized that practically all of these processes are presently used in practice, but some were included which are now in the pilot-plant stage. The list of those possibilities which have successfully been tried in the

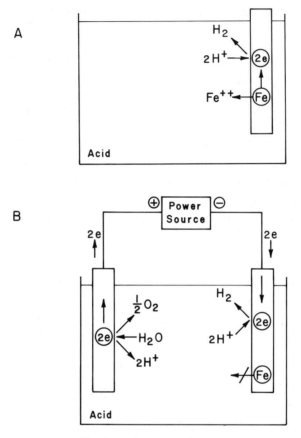

Fig. 9.4. Electrochemical corrosion prevention. (A) A piece of iron is corroding
in an acid solution with concurrent hydrogen evolution. (B) Electrons for the
hydrogen evolution are supplied from
outside, and the corrosion stops.

laboratory but not yet translated into industrial practice would
be much longer. There is much room for engineering development of these, and for discovery of more practically useful
electrode reactions. Ecologically, the importance of many of
these processes is that they do not pollute the atmosphere or
water with side or rejection products.

CHAPTER 10

Electrochemical Pollution Analysis

10.1. Ecological Problems and Analysis

The recent emergence of ecological interest resulted in quite a challenge for the analytical chemist. It is easy to say: we want clean air, but in practical life, one has to be more specific. The harmful impurities have to be exactly identified and a concentration limit must be established, above which their presence cannot be tolerated. A short digression about the myth of "zero pollution" is timely at this point. The absolute removal of "pollutants" is not only economically impossible (some discussion of this was included in Chapter 7), but may also be definitely harmful in some cases. Copper, for example, can be toxic to aquatic life, but it has been shown that a small trace of it is not only beneficial but absolutely necessary. In waters completely devoid of copper, many species cannot live, severely upsetting the food chain of the aquatic life. The same is true for other "pollutants." Rather than talking about "zero" contamination, one should strive for the "optimum" concentration of the chemical in question. Furthermore, this optimum concentration is usually not a fixed number, but a rather wide range; while life thrives in this range, any deviation (*increase or decrease*) will be harmful to life. If one also considers that the removal of the last 1 % of an impurity may cost ten times as much as did the removal of the first 99%, it seems foolish (and sometimes harmful) to try to remove this last 1%, unless it is still on the high side of the optimum. Now, the identification of pollutants, and the determination of the optimum concentration range, involves chemical analysis. That is not all, however; after limits for pollutants have been agreed to and laws have been passed to assure the needed purity, there are going to be no useful results unless compliance with the law can be checked. The antipollution laws cannot be enforced unless a continuous check is kept on the level of pollutants both in the air and in the waters. This requires further analytical work.

Why is pollution analysis a special challenge to the analyst? It is so for two reasons. First, the amounts to be detected are in many cases very small. The tolerance limits of most materials are set in units called ppm (one part per *million* parts of water or air), and in cases of highly toxic materials in ppb (parts per *billion*). One does not have to be a trained analytical chemist to realize that we are talking about extremely tiny amounts and that dealing with such minute quantities will cause extreme difficulties. The second reason is the large number of analyses to be carried out. To quickly detect and pinpoint the source of dangerous pollutants, a number of analytical stations are needed alongside a river where frequent samples can be taken (around the clock) and analyzed for many constituents. Laboratory techniques for water analysis are available to do the job, but building these stations, equipping them with the needed instruments, and staffing them with trained people to do the analyses would be expensive. What is needed is automation: water quality monitoring stations which will automatically take samples and carry out the analysis without the attention of operators and with only infrequent servicing needs. The data then can be transmitted from all these stations by telephone or radio to a central control area where they can be computerized for quick analysis and future reference. The same is true for keeping a check on the atmospheric pollution over a large city. Such a network of stations can give instant warning of pollution incidents, thereby making it possible to take countermeasures in due time and avoid hazardous situations. The development of automated analytical instrumentation and auxiliary equipment is a task for the analyst and the instrument man.

10.2. Electrochemical Analysis

Electroanalytical chemistry is one of the oldest, classical parts of electrochemistry. It is a well developed, mature science; its theory and practice are the subject of many books. Some electrochemical instrumentation is part of every analytical laboratory, and one can state without exaggeration that there are hardly any materials for which an electroanalytical technique would not be available. This, in itself, would assure a role for electroanalytical chemistry in pollution analysis. There is, however, a special characteristic which makes its use more widespread than its expected share. This is due to the fact, as mentioned above, that automation and telemetering of the results are necessities in most cases of pollution monitoring. Most of the automated instruments work electrically and telemetering is done exclusively by electrical means, whether through wire or through air. The results of "normal" chemical analysis are in the change of a color or a color intensity, or they appear as the weight of a solid or the volume of a liquid or a gas. These results have to be translated into electrical signals for automation and data collection. The built-in advantage of the electrochemical methods is that the results appear as electrical signals *by the nature of the methods.* No conversion is needed, with its inherent loss of accuracy and complication of equipment. Of course, this characteristic alone is not enough to justify the use of a method; accuracy and specificity (being free of interference from other components of the sample) are equally important, but the electrochemical techniques always have a head start over other methods in the competition.

Electrical measurement is a broad term and can be broken up into several smaller areas. Just to take some of the most well known quantities: one can measure the current, the voltage, the resistance, or the amount of the electrical charge. It is convenient to classify the electroanalytical techniques according to the measured electrical parameter. We are not going to give a comprehensive discussion of electroanalytical chemistry, we only wish to describe in general terms the most commonly used techniques and illustrate, with some examples, their environmental uses.

Electroanalytical determinations are carried out in the solution phase and their application to water analysis is straightforward. They are, however, applicable also to the analysis of atmospheric constituents after the air sample has been bubbled through some solution to dissolve the desired pollutant.

10.3. Potentiometry

In these techniques, the potential (voltage) is the measured quantity. How can a voltage be a measure of some impurity concentration in a river? The theory is lengthy and will be avoided here; a brief description is given in Appendixes I and II, and specifically the Nernst equation given in Section A2.7 is applicable. According to Nernst, if one has some copper ions in water and dips into this solution a copper wire, the potential of this electrode, as measured against a fixed reference electrode (Appendix I), is a measure of the copper ion concentration. A simple device, two electrodes, dipping into the water and connected to a voltmeter (or to a telemetering device transmitting the voltage signal to a distance control center). Unfortunately, while such a device works perfectly well in the laboratory, the copper wire dipped into a river is useless. First of all, there are many possible side reactions due to other impurities in the water and, secondly, the electrode will be covered with dirt and slime in a short time. The problem, however, is not unsolvable and useful systems have been worked out which can be used in natural waters. The specific engineering and chemical details would be out of place here; the basis of the method remains the Nernst equation described in Appendix II.

One of the classical examples is the measurement of the acidity or alkalinity of the water (the so-called pH). Electrodes for this purpose have been available for nearly half a century. Using another electrode, the oxidizing or reducing power of the water (ORP) can be determined. Both of these parameters are nonspecific; they are not related to a given component of the water since many compounds may affect their value. They are still useful and can be correlated with water quality. More *specific* electrodes (such as for the measurement of copper, as mentioned above) have only recently been developed and are available commercially for less than ten years. There has been fast progress made in this area, and a leading manufacturer offers 18 different kinds of electrodes. The specificity of the electrodes is constantly improved and new ones appear on the market quite fast; the above number will probably be outdated by the time this book is printed. These electrodes are serving mainly for metal ions and inorganic anions, but some have also been suggested for organic materials.

In addition to direct concentration determination with potentiometry, there are also indirect applications. For example, the concentration of a material in solution can be determined from the amount of reagent needed to precipitate it. Let us say we have chloride ions in the water and the concentration is determined by the amount of silver ions which we have to add to precipitate all the chloride, as insoluble silver chloride. We add the silver solution of known concentration with a constant-rate pump (in an automatic setup) and so the *time* needed to feed the needed silver is a measure of the chloride ion concentration. But how to determine when we add enough silver? Here is where potentiometry comes in. We put in an electrode sensitive to silver ions. As long as there is chloride present, the concentration of silver will be practically zero since it is immediately precipitated as soon as it is added to the sample. The minute the first excess drop of silver solution is added and the silver concentration increases, the potential of the silver-sensitive electrode suddenly changes. This signal can be used to stop the feed pump and the timer. (The whole measuring system is more complex and includes a pump to supply a known volume of sample, a control mechanism which periodically repeats the whole process of sample taking, starting the reagent pump and the timer, and relays the time read-out at the end of the cycle.)

10.4. Conductometry

Another electrical parameter we can measure is the resistance (or its reciprocal: the conductance) of the solution. The resistance of pure water is very high (it is a good insulator), but as salts (or ion-containing impurities) are dissolved in the water, the resistance decreases and this change of resistance is proportional to the total amount of dissolved salts. This is again nonspecific (we do not know from the results what kind of salts are in the water), but is a useful and widely used measure of overall water quality.

How do you measure resistance? Put two electrodes in the solution, apply a certain amount of constant A.C. current (I), and from the voltage (V) needed to push the current through, the resistance (R) can be calculated using Ohm's law: $R = V/I$. As the electrodes are left in the solution, dirt will accumulate on their surface and will cause an increase in the voltage needed, thus falsifying the results. A simple way out of this dilemma is to use a four-electrode setup as shown in Fig. 10.1. The two outer electrodes carry the constant current and the two inner ones pick up the voltage drop in the solution. As slime builds up on the electrodes, there will be an increased voltage needed on the two outer electrodes, but the voltage drop measured by the inner electrodes will be unchanged (assuming that the solution resistance is unchanged) since they do not carry significant current and, therefore, are not affected by the coating. This principle is being successfully used in practice.

Occasionally, the conductometric method can be specific, usually coupled with a specific reagent. In this case, the component to be determined reacts with the reagent and the resulting solution has a higher or lower conductivity, the change being proportional to the concentration of ions introduced. For example, water which does not contain oxygen does not corrode thallium metal, but a small amount of oxygen will cause the

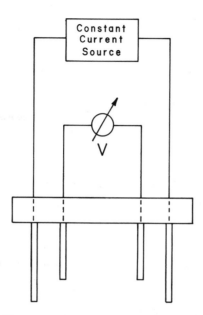

Fig. 10.1. Schematic of a four-electrode
conductivity sensor.

thallium to dissolve, causing an increase in the solution conductivity. Two conductivity cells, one placed before and one
after a thallium-chip-filled column, are then a sensor for dissolved oxygen. Another application of this principle is the
determination of ozone in the atmosphere; the air is bubbled
through a specific solution whose conductivity changes due to
a reaction with the ozone.

10.5. Amperometry

As the name of the method indicates, the measured parameter is the current (amperes). How is current related to concentration? In Appendix III (Section A3.2), we introduce the
idea of limiting current and give an equation for it. The limiting
current is directly proportional to the concentration. It is
therefore a convenient, and easily automated, way of analysis.
There is even a way for separately determining components of
a complex mixture. In Section 8.2, we have discussed that

from a solution containing a mixture of two metal salts, at a given potential only one of the metals will deposit on the negative electrode and an increased potential is needed to deposit both metals. Arranging the experimental setup in such a way as to assure limiting current conditions and slowly changing the potential, a current trace as shown in Fig. 10.2 can, for example, be obtained. At potential V_A, only one of the metal deposits and the current is proportional to its concentration. At potential V_B, both metals deposit and the current proportional to each component can be determined as shown in Fig. 10.2. The method based on this principle is called polarography. It is a widely used, successful technique; the books and papers written about it could fill a small library and a detailed description can be found in every book on analysis. We are not going to discuss it any further except to mention that it is not only a laboratory technique, but has also been adapted for field use.

A special application can be mentioned as an example of interference elimination. Sometimes varying the potential is not sufficient to achieve separation since some components

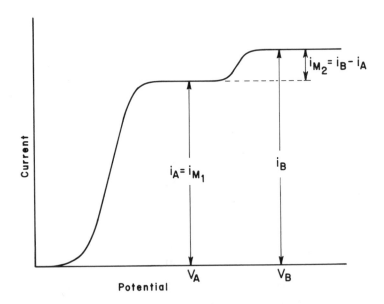

Fig. 10.2. An example of polarography.

may react at the electrode at the same, or at very close, potentials. If, however, a selective diffusion layer can be applied over the electrode which lets only one component through, a perfect separation can be achieved. For example, for the determination of dissolved oxygen in water, a thin plastic film is applied over the electrode; only the oxygen, and not the dissolved salts, will penetrate the film. The potential is not varied (in contrast to polarography), but is kept at a value needed to react the oxygen; the current is directly proportional to the oxygen concentration and practically no interference exists. It is not accidental that this is the second method that we discuss for oxygen determination. Oxygen is a very important water quality indicator and aquatic life depends on its presence.

Other ways to utilize current measurements for concentration determination also are used, but these suffice as examples.

10.6. Coulometry

We already used the example of chloride ion determination (an always present, and often analyzed, constituent of water) in Section 10.3. There we used a pump to introduce the silver reagent solution. An electrochemical method could also be used: a silver electrode is placed into the solution (see Fig. 10.3) and is dissolved with a constant current, thereby introducing the silver ion reagent into the sample at a constant rate. The rest of the setup is the same as before: a silver-ion-sensitive electrode senses the excess silver and, when all the chloride has been precipitated, stops the current and the clock. The time reading is proportional to the chloride ion concentration.* Automatic analyzers based on this principle are commercially available.

---

* Actually what we measure is the electric charge (coulombs) needed to introduce the required amount of silver reagent (current times elapsed time is charge, and since the current is constant, the time is proportional to the charge); and since a known amount of charge is needed to dissolve one gram of silver (cf. Faraday number in Section A2.5), this gives the wanted result.

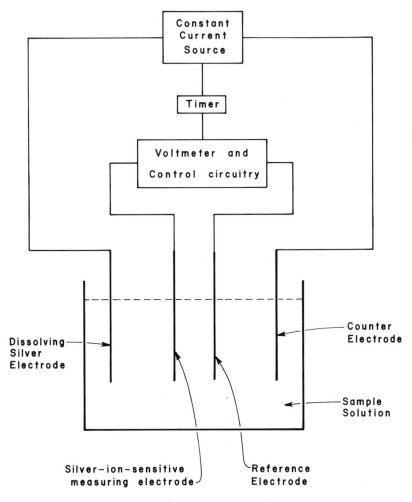

Fig. 10.3. Schematic of coulometric chloride determination.

This is only one example of the electrochemical generation of reagents. Another way to apply the coulometric principle is to react the material in question at an electrode and measure the needed charge.

10.7. Summary

We have hardly scratched the surface of the science called electroanalytical chemistry; our goal was not to enable the reader to make his own air or water analysis stations, but only to show the applicability of electroanalytical techniques to the problem. These techniques are widely used in practice, both in the laboratory and in field monitoring. As an example of the latter, Fig. 10.4 shows a photograph of an automatic water quality monitoring station, one of several keeping a continuous watch on the river waters near Philadelphia.

Fig. 10.4. Pollution monitor designed to rigid federal specifications. Engineer examines a pH sensor, one of twelve different sensors used to make water quality measurements. (Photo courtesy Honeywell, Inc.)

CHAPTER 11

The Electrochemical Future

11.1. Introduction

The future of technological man in the early 1970's looks bleak. Just at the moment when he has so much going for him, and when an end to compulsory work for most of the population (abundant electricity from solar or atomic sources, plus a programmed control of industrial processes) appears as a real possibility not in some remote future, but perhaps for one's children's children, he has developed a case of future shock. He wants to eliminate much of the science and technology which are his main weapons in dealing with the technological world to which he is now so tightly connected. The fact that this turn of mood has come upon the most technologically powerful nation, to which country so many others look for technological leadership, is a misfortune the damage from which will not be confined to the United States. The point is that the *present* technology, *coupled* to the present population, disbalances natural processes and exhausts materials.

In speculating in this chapter on the next 100 years, an entirely serious speculation is that there won't be a future for technological man, that we shall fail one of the series of tests we face between now and 2050, and that some population-reducing calamity will be the result, so that our problems will be solved, just as illness can be solved by death.

However, if something more acceptable may occur, then it is interesting to pick out those parts of a possible favorable future which seem likely, and those parts which can be argued. There seem to be three points which are effectively certain.

(i)   The burning of fossil fuels will cease to be the main way we get our energy early in the next century.

(ii) The new sources of energy will be at first the fission of nuclei, and then solar energy and perhaps fusion.

(iii) Electricity is likely to be abundant by about 2025 A.D. (it could be abundant a generation earlier, but there is not at present the political will which would make the necessary research funding possible). "Abundant" means: about one-fifth the price in present dollars, and about five times as much consumed per head of the population.

The rest is a matter of attempting to discern probabilities. A considerable part of the future will involve electrochemical technology. The thesis of this book is that technological affluence and ecological balance can be maintained by an extensive development of electrochemical technology. Thus:

(a) The medium of energy will be electricity: this will be abundant and cheap.

(b) Some central parts of future technology will be based upon electrochemistry, or electrochemically derived products, e.g., transportation.

(c) In a world in which no pollutant may enter the atmosphere or waters, and in which no fresh material enters, many things—eventually all metals—will have to be recycled; and in a world in which electricity is abundant, the ecological aspects of the recycling processes will be met best by the use of pollution-free electrochemical processes.

11.2. The Hydrogen Economy

One very likely aspect of the electrochemical future is that, instead of reactors or solar farms producing electricity to be transmitted through wires, the electricity will be applied at site to produce hydrogen electrochemically from seawater. As a major environmental difficulty with big reactors is the *thermal* pollution they will cause, it would seem reasonable to have them out at sea, so that the thermal pollution problem is solved by heat rejection to the oceans.

The product of these energy islands would be hydrogen, and it would be piped back to land, perhaps 10–50 miles distant,* and thence to the cities. In Fig. 6.1 we have shown the advantage of doing this, costwise, compared with sending electricity through wires.

The hydrogen arriving at factory and home will be converted to electricity by hydrogen-air fuel cells to the extent that energy is needed in that form. Pure water will be generated as a by-product of this process. For heat the hydrogen will be burned directly.

Some advantages of this are as follows:

(i) It will give electricity more cheaply than if transported through wires.
(ii) It will solve some problems of the availability of fresh water, because the electricity expected by 2000 A.D. per person will produce (if supplied by means of hydrogen) the drinking and cooking water needed per person. Habitation of arid regions would thus be made easier.
(iii) It will give us an energy system which has *zero* pollution, so that an increase in world energy production to a point at which the present poor countries could join the wealthy ones would be possible.

---

* It could be that the islands are very much further off if they are solar collecting platforms. Oil pipelines stretch many thousands of miles over land, and piping hydrogen over long ocean passages does not seem an unlikely prospect.

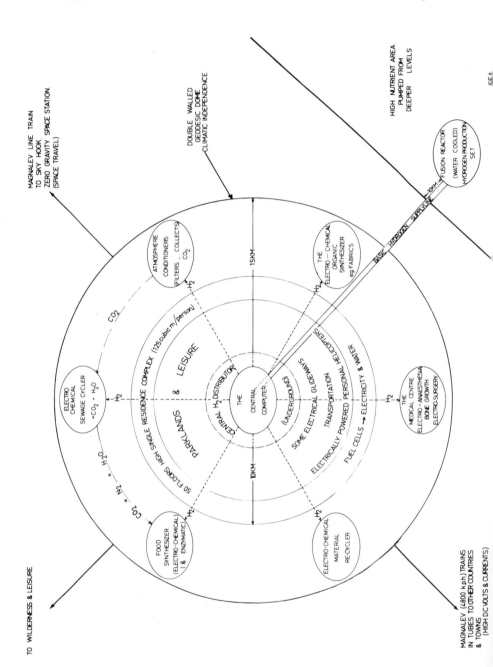

Fig. 11.1. Town of one million population (2050). This figure shows a possible scheme for housing about one million people in an advanced application of the hydrogen economy. A reactor floats on the sea, electrolyzing water, and all the functions within the town are electrically driven with electricity obtained from hydrogen reconverted to electricity by means of fuel cells. The town is self-contained, the food and material are cycled. No actual work has to be done by any person, except technicians to repair the computer-controlled, electrically driven, factories.

The advantages compared with sending electricity through wires are: the provision of water; reduced costs of power; no unsightly grid lines; the production of D.C. power (needed to drive electrical industry, magnetically levitated vehicles); the provision of very cheap excess hydrogen for chemical synthesis (the direct reduction of iron ore, etc.).

Very cheap oxygen would also be available.

11.3. Future Towns

Two things are clear about the towns in the era after 2000 A.D.:

(i) They must be built upon ecologically sound principles for the long term, not with 100, but with 1000 years in view. Essentially, this means that practically all materials utilized by mankind would have to be recycled artificial-

(ii) The concept of compulsory work by individuals to maintain their livelihood will die because of the availability of abundant energy and automation. The only work which will *have* to be done will be that of maintenance of the automated processes. (This doesn't mean that people won't work. They merely won't have to work to be able to live affluently.)

We could be near to the realization of concepts such as these, not more than two or three generations away in developed countries. It is principally a matter of the funding of research to bring about sufficient energy per head. However, these considerations neglect the psychological factors which may slow down the changeover to a postindustrial society.

A sketch of one set of options for a future town with some electrical and electrochemical features is given in Fig. 11.1. Energy, transportation, heating, recycling, purification, and the production of food and organic chemical products (as well as some of the medical treatment) will involve electrochemical processes, some of them through the medium of hydrogen.

In the diagram, nothing is suggested beyond present principles, and what fundamental research indicates *should* be developed; much needs applied research and new engineering.

11.4. The Type of Research Needed to Develop These Concepts

Education, from high school upward, in the principles of electrochemical science is implied as a basic step in getting people to realize the possibilities. With respect to training at the university level, we certainly need theoretical and fundamental electrochemists, but we also need, and in far greater numbers, electrochemical *engineers*, with a knowledge of *modern* electrochemistry. Furthermore, many other "interdisciplinary types," such as materials scientists and energy conversion engineers, will be needed to research and develop the necessary electrochemical technology.

The main things which seem to stand between man and the development of a society in which he will no longer have to fear ecological catastrophies and/or a stunted future is a lack of knowledge of the scientific possibilities, a lack of initiative and will, an inertia of the status quo in the capital- and initiative-controlling organizations in the United States. World-wide there is a lack of awareness of the seriousness of the situation. Such an awareness would make the control of one of the greatest dangers—the growth of the population—politically acceptable.

11.5. Ecology and Electrochemistry

Ecology is a broad discipline. It involves not only the sciences and engineering, but, e.g., law and medicine.

Although it is true that a satisfactory arrangement of both a high living standard and the large population to give a restored ecological balance will take all the subjects mentioned, and many more, there are certain core sciences within ecology, including demography, energy conversion, chemical technology, automation. It is through the relating of energy to chemical technology that electrochemistry arises as a core science for the ecologist.

In the future, energy will be available chiefly in the form of electricity. Electrochemistry is the science which joins energy to materials, which allows the prospect of a pollution-free electrically powered technology using continuously recycled materials.

Lastly, it never helps to oversell a field. We wish to be realistic in our enthusiastic presentation of electrochemical science as a core science for the development of new technologies. This is a big claim, and we reiterate it. But we don't mean that electrochemistry will be the basis of material aspects of the future to the exclusion of chemistry.

Many antipollutant measures and some recycling may be carried out better chemically. For example, the achievement of a hydrogen economy seems a likely prospect. The hydrogen will probably be produced by electrochemical means. But, whether it will be *used* to make electricity in a fuel cell, and then this electricity used in an electrochemical method, or applied directly in a chemical way is a matter which would have to be considered for each problem. For example, it is attractive to think of liquid hydrogen just replacing petroleum in transportation. Powered in this way, cars would still be massive, made of great quantities of steel. They would be much more expensive to run than those run electrochemically, but this could perhaps be concealed or disputed. Without the funding of the necessary research, the silent, long-lasting fuel cell electric cars would never be so fast and powerful as those driven by hydrogen combustion motors. And the companies would sell more "gas" for the same number of miles traveled.

The main pulse behind this book is to give information to managers and engineers about a new field and its possibilities, for those possibilities seem fitted and acceptable for a permanent nonpolluting technology. But, again, we must balance with the negative side. A manager who wants to set up a device involving a large amount of high-pressure air—to work, say, a pneumatic drill—can phone in an order and have delivered promptly a diesel-oil-burning behemoth which will smoke, thunder, vibrate, and roar across his plant for the time of use. *If* he (and that may be a rather big if) *knows* enough electrochemistry, he can work out how to compress air completely silently and cleanly to effectively any pressure by electrochemical means, and it won't be difficult for him to figure out how that high pressure could be transduced to his drill (the sharpness and effectiveness of which could be electrochemically controlled). But he cannot rent or buy one. It is not yet manufactured. The possibility of the application has only recently been realized.

But the prospects look even worse if this manager is a persistent fellow, and persuasive. He wants to fund research on it, but he will find that for every electrochemical researcher there are about one hundred chemical ones. Unless he understands very clearly enough about the uniqueness of many of the possibilities to look further and find the few little research companies which exist and the *one or two* which have personnel with modern knowledge, he will probably just stand the noise, smoke, vibration, and thunder of the old way.

To help break out of this dilemma, to show what can be done, and also what *could* be done—that is why we wrote this book.

The Structure of the Region between Phases

A1.1. The Electrical Character of Interfaces

This appendix is about the molecular-scaled stage on which the electrochemical play occurs. It is a region between a phase of the solution, a liquid containing charged particles (ions), and another phase in which the principal mode of transport of charges is by the motion of electrons. Normally, this situation would be between a metal, for example, nickel, and a solution, for example, seawater.

Our main interest in this interfacial region is in the electrical characteristics: the free excess charges existing on each side of the interface, the description of the potential differences which exist between the two phases, and the strong electric field which this potential difference (a difference of about a volt in a few angstroms) causes.

This interfacial region is often called the *interphase,* to emphasize the three-dimensional character of the situation. When two phases (say, a liquid and a solid) are brought into contact, their interaction will create what we may consider a third phase between them. The presence of the solid in contact with the liquid will change the characteristics of the liquid from that observable in the bulk phase, and this change may penetrate many hundreds of angstroms into the liquid; a similar effect may also exist in the solid side of the interface. Therefore, if we consider the "surface" as the layer of the material where the characteristics are different from those of the "normal" (bulk) values, we have to think about a finite thickness, rather than a two-dimensional (geometrical) surface; hence the name inter*phase.*

A1.2. The Origin of the Electrification of Interfaces

It is not difficult to understand why all interfaces are charged. Thus, if we consider a metal and focus our attention on its bulk, then the positive nuclei of the metal ions there and the negative electrons are equal in number, and there can be no net charge inside the bulk. But suppose we make a thought experiment in which, in some way we do not need to define, we can cut through the metal instantaneously. Suddenly, the electrons on the instantaneously produced surface are not in the same situation as they were before. The forces on them differ in the one dirrection. For this reason, they will tend to pour out, leaving an excess positive charge on the metal, and this positive charge will build up until it has produced a counterforce on the outgoing negative electrons sufficient to prevent their further exiting. A new equilibrium will exist.

Some argument such as this would always apply to a newly formed surface, whatever its composition. Interfaces are always electrified, and can be negatively or positively charged, depending on the circumstances. It is to be emphasized that while the interfaces are charged, the *net* charge in the interfacial *region* (the interphase) is zero, that is, the sum of the charges on the two sides adds up to zero.

A1.3. There Is Always a Potential Difference Across Every Interphase

There is not much need to argue this, after we have shown that each interface will have a net charge upon it. For if one interface has a net positive charge and the other a net negative charge, there will obviously be a difference of electrical potential between them.

A disappointment to fundamental electrochemists is that there is no way by which they can *measure* one of these potential differences. They do know that the order of magnitude of interfacial potential differences is usually in the region of a few hundred millivolts, but they have not yet discovered a method by which they can actually measure it.

Here is why: look at Fig. A1.1. If we want to measure the potential difference across the interphase formed by a metal, *M,* in a solution, *S,* we can try to do it by forming another interphase in the same solution, say, that between another metal, *N,* and the same solution, *S,* and join both to a voltmeter. But, of course, this only measures the *difference* in the potential differences between the interphase at *M* and *S* and the interphase across the situation *N* and *S.* We are none the wiser about either of the potential differences across these interphases *individually.*

Many ideas have been expounded which are aimed at measuring the single potential difference across an interphase, but none has succeeded.

A1.4. A Change of Potential at an Interphase Which Can Be Experimentally Measured

Although we have just asserted that it is not possible to measure absolutely the potential difference across one interphase, it *is* possible to measure its *change.* We do it as follows.

Consider the setup shown in Fig. A1.2. We measure the potential difference between the electrode *M* and another which for reasons which will become clearer in a moment we will call *"Ref."*

Suppose that we have arranged things so that this measuring device is adjusted so that there is zero current flowing between the electrode *M* and electrode *Ref.* Then, the potential which this voltmeter shows must be the difference between the potential difference of the interphase between *M* and the

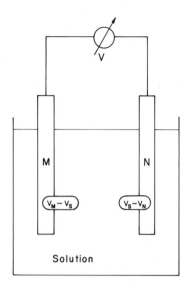

Fig. A1.1. Attempt to measure the metal–solution potential difference. Note that when one "connects" the voltmeter to the solution, a second metal–solution interphase is created unavoidably.

solution and that of the interphase between *Ref* and the solution.

Suppose now that we attach the electrode *M* to another circuit, as shown in Fig. A1.2; we then inject charges into the electrode *M,* or take them out. We can make the excess charge on the metal *M* more positive, or more negative, as we want, depending on the polarity of the power source. As *M* is one side of the interphase, if we change the charge across the layer, the potential difference across it will change.

The point of this section is, can we measure this change? The answer is, yes indeed. If it were not yes, there would not be much electrochemical science. The reason we can measure the *change* (though not the absolute value) can be seen from the circuit. This, in this circuit the potential at *Ref*—or at the boundary of *Ref* with the solution—is constant, while we

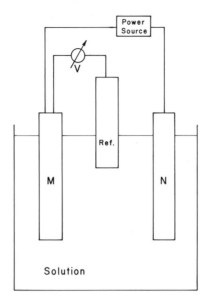

Fig. A1.2. Measurement of the potential change with the use of a reference electrode.

change the potential at the M–solution interphase. (Note that there is no current flowing through the Ref–solution boundary so there is no disturbance of the original equilibrium). Therefore, if we push more charge, or less of it, into M, the voltmeter will measure the *change* of potential. It will look as though the potential of the cell M/S–voltmeter–Ref/S has changed; as we have seen that at the interphase Ref–solution there is *no* change, the *change* observed in the measured potential is just that of the change of the unknown potential at M–solution.

A1.5. Scale of Potential Difference

To have some kind of communication system for potentials, all we have to do is to make sure that we always have the same reference system as we change the other electrode. Then, if we measure a potential which gives rise to a value on the voltmeter, with a certain $M-S$, and then bring in another system, we can determine a voltage difference. For example, $M-S$ is silver in contact with silver nitrate solution in the first case, and copper in contact with a cupric nitrate solution in the second measurement. From the two voltmeter readings we can determine that the potential of silver with respect to silver nitrate solution is 0.46 V more positive than the potential of copper with respect to cupric nitrate solution,* though, at the same time, we have no idea what the silver–silver nitrate solution or the copper–copper nitrate solution potential differences themselves are.

If we now select a reference electrode and arbitrarily assign its potential a value of zero, we have a relative scale on which to compare potential differences and with the help of which calculations can be carried out. There are numerous requirements for a suitable reference electrode, but a detailed discussion of these would not give any further insight as far as the basic idea is concerned. Let it suffice to say that such an arbitrary zero, and the reference system to define it, has been selected by international agreement.

While such a scale is completely arbitrary, it is quite useful for two reasons:

(i) Using this scale, one can predict from handbook values what the potential difference between two pieces of metal dipped into a given solution will be. Of course, this will be only the *difference* between the two metal–solution potential differences, as described above, but then, this is the only thing measurable, anyhow.

(ii) The current–potential relations (to be discussed in Appendix II) which form the basis of modern electrochemical science can be described in terms of *changes* of the electrode potential, rather than in its absolute value. Therefore, while the lack of the *precise* knowledge (rough estimates can be made from theory) of electrode–solution potential difference is disappointing in many cases, electrochemistry still has a sound, quantitative basis.

---

* The numerical value is true for certain "standard" conditions only: unit activity of the solutions, and 25°C.

A1.6. Adsorption at Metal–Solution Interphases

Adsorption of atoms in the gas phase on a metal is simple to describe. It is an on–off matter, very well defined. When the gas molecules become extremely close to the metal, perhaps three or four molecular layers from it, they feel the properties of the metal, and, suddenly, cling to it. Thus, adsorption at a metal–gas interface is effectively measured by the coverage of the metal with the adsorbed material, and we don't really have to think of any other type of change of concentration in the gas phase.

Adsorption at a metal–solution interphase is a different situation from that of metal–gas interface because the electric fields which spread out from a metal into a solution often cause significant concentration changes at distances of hundreds of angstroms from the electrode.

An analogy would be the effect of a gently sloping shore upon the depth of water (the concentration–distance profile in the metal–solution situation), compared with that of a sharp brick wall (the concentration–distance profile in the metal–gas situation). This is interphase vs. interface.

Thus, in Fig. A1.3 the concentration of ions is shown as a function of distance as the metal–solution interphase is approached.

To emphasize this difference from "normal" adsorption, the change in numbers of ions per square centimeter of surface from that which would have been there in the absence of an electrode is given a name: surface excess. Figure A1.4 illustrates the surface excess of the negative ions of the situation of Fig. A1.3. Note two important characteristics: this considers not only the interface but the whole interphase; secondly, it is not the total number of ions present in the interphase but the excess over and above what would be present in a similar volume of solution in the absence of the interface. From this, it follows that the surface excess could be positive or negative (surface deficiency, if you will), e.g., for the positive ions of Fig. A1.3 the value is negative since their concentration is lower in the interphase than in the bulk. It is exactly this difference of the surface

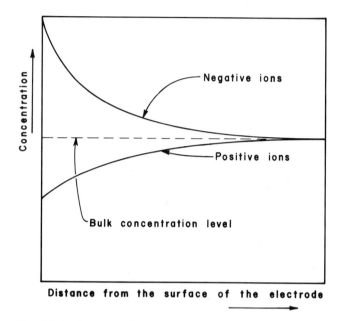

Distance from the surface of the electrode

Fig. A1.3. Concentration distribution of the negative and positive ions at a metal–solution interphase. In this case, the excess charge on the metal is positive, and that of the solution is negative.

excesses of the positive and negative ions which causes the net charge on the solution side of the interphase.

We may say that in certain limiting cases, when the solution is very dilute (so that the amount present near the surface is small) and the change in concentration caused outside the electrode is small, the surface excess is nearly the same thing as adsorption (which is just the number of particles actually in contact with the substrate per square centimeter).

A1.7 The Structure of the Interphase

In the preceding section, we have indicated that a metal–solution "interface" is more complex than a metal–gas situation. The first picture we have given is an oversimplification:

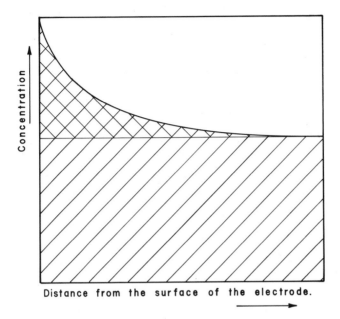

Fig. A1.4. Concept of surface excess. The total area under the
curve represents the amount of ions at the surface. The single-
hatched area is that of the normal (bulk solution) value; the
double-hatched area is the *excess*.

there is more structure to the interphase than we have shown.
What we mean by structure at an interphase is the varying
positions of ions in solution, starting out from the electrode.
For example, we find that water molecules from a solution
are attached to the electrode, oriented in a certain direction
perhaps, and that this orientation changes with potential.
We may find that chloride ions in a sodium chloride solution
are attached to the electrode at such and such a concentration,
but that some of them are distributed into the solution, according
to a concentration distribution law which varies with distance.
Correspondingly, we find that sodium ions are distributed
according to another expontial law, and so forth.

The build-up of information of this kind is called the
"study of the structure of the double layer," and the structure
that is meant is the variation with distance of the distribution

of ions and molecules out from the surface into the solution.

Most of the structure of the interphase is 5–10 A from the metal surface, but there is some structure—that is to say, some different distribution of cations and anions—further out into the solution than that, and if we take dilute solutions, say solutions of concentration $10^{-4}M$, there are differences in concentration as far out as 1000 A.

Thus, we find that the positively and negatively charged ions (cations and anions) do not behave similarly.

(i) Cations—at least small ordinary cations such as sodium —do not adsorb right on (in contact with) a metal. Cations are well dealt with by taking them as just electrostatically influenced by the presence of the surface, but basically outside the contact with the metal, giving the type of distribution shown in Fig. A1.3. The reason is that cations are mostly heavily bound by water sheaths, and therefore their ion centers never do get near the metal.

(ii) For anions, however, with exceptions, there is binding of a special kind to the surface. This binding is not a chemical bond, but some specific interaction between an anion and a metal makes a portion of the anions bond more closely to the metal, while the rest of them are electrostatically distributed outside the metal, just like the cations.

Then, we have to consider the presence of the solvent (usually water). If we plunge into an aqueous solution—which consists of dipoles of water bonded together in various ways—a metal plate with a high field strength at it, the dipoles of the water will not continue bonded, but will be broken up. Some of them will stick to the electrode, and depending on the sign of its charge, water molecules will stick with their hydrogen away from the electrode or with their oxygen away from the electrode.

Further, we have to consider the effect of organic molecules added accidentally or purposefully to the solution. Here, there are also two kinds of situations to consider. We are only going to consider one of them, however, that in which organic molecules adsorb on electrodes without dissociation. There is another, more complicated case, in which organic molecules adsorb with dissociation.

Consider a situation in which we have an idealized organic molecule which does not react with the electrode, does not dissociate, and does not have any polar groups, so that it itself will not interact with the electrode field. It might be a fairly long-chain hydrocarbon, such as decane.

Now, come back to the picture drawn above of water molecules on the electrode. If the hydrocarbon adsorbs on the electrode, it has to displace the water molecules. In the limiting case, where the organic does not react with the electrode, the influence of potential on the adsorption of the organic will operate only through the effect of potential on the water adsorption. Suppose that we are at an extremely negative potential with our electrode. Then, the water molecules will be held strongly, with their hydrogen atoms oriented toward the surface of the electrode. Under such conditions, the organic will be displaced from the surface. The same will happen when the electrode tends to be positively charged; here the water molecules will be reversed in their adsorption on the surface, and the oxygen atoms being strongly attached to the metal. When the electrode charge tends to zero, there will be a minimal strenght of attachment of the water molecules to the surface, and the organic molecules will have an easier job of displacing them, and will adsorb. Something like a parabolic curve can at once be seen to result from this model and, indeed, experimental measurements indicate that the adsorption of organics goes through a maximum as the potential of the electrode is changed.

There are other complications concerned with the adsorption of organic molecules. When they themselves become highly structured in an electrical sense, with highly polar groups on the surface, the organic–electrode interaction will become strong, perhaps in some cases stronger than that of the water–electrode interaction, and a somewhat different behavior may result. However, it is true that, at least at low concentrations of the organic at the surface, the relation between the coverage and the electrode charge tends to be parabolic.

Another interesting aspect of this situation is the question of the potential at which the maximum adsorption occurs. The discussion given above suggests that it will tend to be at the

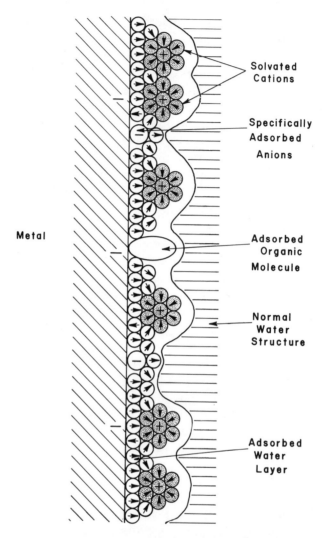

Fig. A1.5. Structure of the interphase in the near vicinity of the metal–solution interface. The concentration changes reaching deeper into the solution (*cf.* Fig. A1.3) are not shown.

potential of zero charge, because there the water molecules are least strongly attached. We find, experimentally, that this is not always correct, especially on solid metals. One reason for this is that the attachment of the water molecules to the surface is not always simple; there may be a greater tendency for the water molecule to be adsorbed in one direction than in the other. Another reason is that the rough model we give above neglects a potential contribution from the organic molecules and that may indeed be there if the molecule has charged groups.

Figure A1.5 attempts to give a model of the whole complex situation, showing the special adsorption of the anions, the layer of adsorbed specially oriented water, and some adsorbed organics.

All the above, of course, applies to the solution part of the interphase. Is there no structure to be mentioned in the solid? As it happens, for most practical situations there is none. There is, of course, excess charge on the metal, too (to complement the excess charge of the solution), but that is limited to a narrow surface layer, without penetration into the bulk of the metal; the situation here approaches the ideal (two-dimensional) surface. While this can be generalized for metals, and the electrochemistry we treat in this book is mainly concerned with this situation, it cannot be generalized for every solid–solution interphase. When the solid is a semiconductor, for example, there is a complex interphase structure also on the solid side. There is no need, however, to complicate the description any further for this book.

A1.8. Some Practical Considerations

What does all this have to do with the pollution of Lake Erie, or the atmosphere over the city of New York? There are both direct and numerous indirect connections. The charge separation at the solid–solution "interface" is directly involved in such practical processes as electroflotation and electroflocculation (Section 7.6), electrofiltration (Section 7.7), and electrocoating (Section 9.5), where the particles suspended in aqueous solutions were influenced (i.e., moved) by the application of an electric field because of their excess charge. Furthermore, all electrochemical reactions, many of which have been shown to be useful for the world of an electrically based clean industry, will take place in interphases with structures described in this appendix, and it is not surprising to find that these reactions can be profoundly influenced by the structure of the interphase. For example, let us take only the adsorption of organics at the electrode surface. This can block the surface of the metal and stop or slow down the charge transfer between the phases. This is sometimes beneficial (it is another electrochemical method of corrosion prevention, the addition of inhibitors to solutions to slow down the destructive dissolution of metals) or it can hinder a desired reaction. If we are considering the reaction of an organic molecule (e.g., electrochemical sewage treatment), we find a double effect: the molecules to be reacted have to adsorb on the surface, but the reaction product should desorb fast to make room for further reaction. It is obvious then that to understand and control when and how these molecules adsorb and desorb is a practically important question in electrochemistry.

In more general terms: the interfacial potential differences discussed in this appendix are the ones which drive the electron transfer through the interface. (This is the essence of an electrochemical reaction.) If we want to control these reactions, to influence their rate or path to achieve a given practical goal (whether that be a waste stream purification process, or the fabrication of some useful product), we have to understand the "surroundings" (the interphase) and their effects upon these processes.

APPENDIX II

Interfacial Charge Transfer

A2.1. Introduction

There are two basic parts in a presentation of modern electrochemistry. One concerns the structure of the double layer and that we have looked at. We might say that Appendix I was a setting up of the main scene in the field.

The other concerns the actual act of charge transfer. We might say that this appendix is putting the actors and the actresses on the stage. How they work, what pulls the strings, that is what concerns us now.

We are going to show how to transfer electrons from an electronically conducting phase to a level in some ion in solution; or to transfer an electron from the ion back to appropriate quantal levels in the solid phase. These things can give rise to a synthesis of chemical compounds, electricity is converted to matter. But we shall also show how we can make things go the other way. Matter can be directly converted to electricity. The two aspects give us the basic steps in how to replenish our world, how to recycle the things which we have used, and how we can obtain electrical energy in a clean fashion.

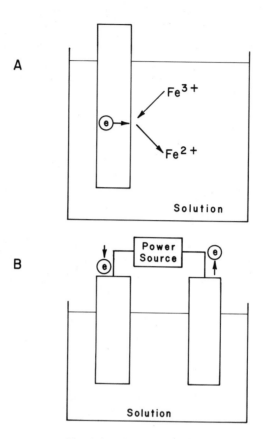

Fig. A2.1. (A) A single-electrode reaction. (B) In an electrolysis cell, such
a reaction can produce useful products
with the consumption of electrical
energy from the power source.

A2.2. Charge Transfer and Its Chemical Implications

Some of the basic ideas of electrochemical reactions were
already mentioned in Section 4.3; they will be restated here in
somewhat more quantitative form. In Fig. A2.1A we show a
metal phase in contact with a solution. The fundamental act
of electrochemistry is the transfer of an electron which exits from

a metal and settles onto an ion in solution. In its simplest, basic form, the electrochemical step can be exemplified by

$$Fe^{3+} + e_M \rightarrow Fe^{2+}$$

where e_M is an electron delivered from the metal to an adjacent ion.

Thus, a ferric ion has been converted to a ferrous ion, a chemical change has been brought about by a piece of electrochemistry. The supersimplicity of the example may deceive. We can cause complicated organic chemical changes, too; for example, we can make the organic substance methanol from the inorganic substance CO_2, selectively introduce different groups into the benzene ring, etc.

A2.3. Charge Transfer and Its Energy Implications

In Fig. A2.2A a metal phase is again shown in a solution, and a hydrogen molecule is about to land on the surface. The molecule lands, dissociates, adsorbs, and injects an electron into the metal. This is the beginning of the process of obtaining electrical energy from an electrochemical change.

What is the difference between the two examples, bringing about a chemical change by putting in electricity, and extracting electricity from a chemical change? It is a matter of how the rest of the circuit is set up. Let us look more into how we can fix up the reaction when the objective is a chemical change, that of the creation of the ferrous ion from the ferric. We can see that it must have some more circuit to it. We cannot just grab electrons out of nowhere and send them into the metal and make them emit. The rest of the circuit is shown in Fig. 2A.1B.

It shows a cell in which electricity causes chemical transformation. We have to have an outside source of power, the electricity source, and this source pushes electricity into what is a reactor (two electrodes, a solution: a cell). In the reactor, the main point is the substances produced. Electricity comes in and forces a chemical reaction to go in a direction in which it will not go spontaneously. That is how we create new matter by electrical currents.

What about the situation in which we get electrical currents (and therefore electrical energy) from chemical substances. What is the full circuit there? It is shown in Fig. A2.2B. It has similarities to the reactor circuit, with one great difference. There is no power source; in its place there is a load. By "a load" we mean, e.g., an electric motor, something which takes in electrical energy and give out mechanical work, a transducer.

Is there any other difference between the two cases, the reactor which produces materials when we push in electricity and the energy generator which works spontaneously, the fuel cell? There is one important difference and that is that the reaction which goes in the solution, which makes the fuel cell work, is going *spontaneously*. It wants to go that way. It delivers its electrons to electrodes; its energy balance is such that it will go spontaneously that way. The final products (for example, for a hydrogen–oxygen fuel cell, water) have less energy than the materials with which we started. On the other hand, with the reactor, where we have to put electricity in, the final products have more chemical energy than the starting materials.

With these two examples, we have shown the basis of how we get new materials (for example, convert the poisonous cyanide of a rejected factory liquor to harmless carbon dioxide) and get energy from materials, for example, electricity from hydrogen and oxygen directly.

Both these examples have direct applications to ecology in waste removal and possible regeneration, and to the *clean* production of electricity and power.

A2.4. There Must Be Two Electrodes in a Cell

We are talking here about an electrode–solution interface, electrons going into the electrode or electrons coming out of it. But that is just as if we were talking about the process of clapping and examining only one hand. There must be *another* hand, too. And there must be two electrodes to form a cell.

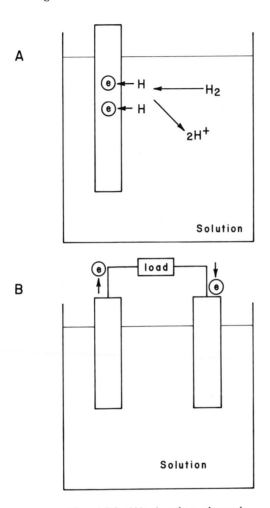

Fig. A2.2. (A) Another electrode
reaction. (B) In a fuel cell, such
a reaction can produce electrical
energy to be utilized to drive a motor
to do, e.g., mechanical work.

The minimum electrochemical contrivance always has two
electrodes, and they connect together through an electronic
conductor. (In Fig. A2.1B and A2.2B, we already have included
two electrodes in the full circuit, without emphasizing that
point.) As we showed in the preceding section, they either

connect through a load, and give out energy, or are driven by some power source and take in energy. The point is that there will be the same current through both electrodes. This means that the same number of electrons *per electrode* per unit time will leave one electrode and go into the solution and leave the solution and go into the other electrode.

For simplicity, one assumes that the two electrodes of a cell have the same surface area, but this need not be so at all. We usually deal in current per square centimeter. We represent this by i, whereas we represent the total current (over the whole electrode) by I. Thus,

$$\frac{I}{A} = i$$

where the A is the area of the electrode.

A2.5. Rate of Reaction and Current Density

There is a simple relation which we should deduce here. It gives a connection between what every physical chemist and chemical engineer knows—the concept of interfacial reaction rate—and what electrochemists deal with—currents and current densities.

Consider the diagram in Fig. A2.2A. It shows an electrode reacting with a hydrogen molecule. If we were dealing with this matter in ordinary chemical kinetics, we could speak of a reaction rate in terms of moles per square centimeter per second. Now, in electrochemistry, we deal with it in different units. We first note that there is a law due to Faraday which states that when we want to deposit one gram-ion at an electrode, and it has a valency of 1, we will have to give 96,500 coulombs to deposit this mole.

The coulomb represents an amount, or quantity, of electricity. It is given by the product of current and time. We might say: coulombs per square centimeter is $i \cdot t$, where i is in amperes per square centimeters and t is in seconds.

Hence,

$$\text{coulombs cm}^{-2} \text{ sec}^{-1} = i$$

Now, if we divide this by the faraday (the coulombs one has to have for 1 mole), we get

$$\text{moles cm}^{-2} \text{ sec}^{-1} = i/F$$

More generally, we would write

$$i = nF \cdot (\text{moles cm}^{-2} \text{ sec}^{-1})$$

Thus, current density is a measure of the reaction rate in moles cm^{-2} sec^{-1}—it is that rate, but multiplied by the constant nF. Here n is the number of faradays which is necessary to make the reaction take place to an extent of one overall reaction amount, e.g., for $2H^+ + 2e \rightarrow H_2$, $n = 2$, while for $Fe^{3+} + e \rightarrow Fe^{2+}$, $n = 1$.

Current density, therefore, is the equivalent in electrochemical kinetics of reaction rate in chemical kinetics.

A2.6. Rate Is a Function of a Potential Difference Across an Interphase

To come to one of *the* bits of fundamental electrochemistry, we must give the relationship between the current density and the potential difference which exists across the interphase, that which we were dealing with when we were talking about double layers in Appendix I.

While the deduction of such a relationship is not too difficult, it is fairly lengthy, and we will not give it here. We can, however, justify the final result, at least roughly, and discuss the consequences in some detail. Even without proof, the relations given in the following will not be in a strange format for readers who are a little familiar with reaction rate concepts in chemical kinetics. In the case of ordinary chemical reactions, the rate depends exponentially on the temperature. Given this, it should not be surprising to find that, in the case of electrode reactions, where the basic step, the transfer of the electron, proceeds under the influence of the interfacial potential difference, the current density (which has been shown to be proportional to the rate of the reaction) is exponentially dependent on the potential.

Let us take a simple electrode reaction, written schematically:

$$A \rightarrow B + e$$

A molecular species, denoted by A, reacts at the electrode, giving up one of its electrons and changing into another species, here called B. The current density can be expressed as

$$i = i_0 \exp\left[(1 - \beta)\frac{F}{RT}\eta\right]$$

This simple expression contains quite a bit of basic electrochemistry. Let us take a closer look at some of the terms. One of the most important features is the appearance of η, the "overpotential." As we have shown in Appendix I, the actual potential difference between the metal and the solution is an unmeasurable quantity, and any relation containing it would be useless for practical purposes. The overpotential, however, is a *change* in the metal–solution potential difference, and as such it is quite easily measurable. It is the shift in the metal–solution potential difference which occurs when the current is changed from zero to some definite current density, i.

Let us emphasize this: the overpotential is the change which we *must* make in the metal–solution potential difference, starting from its equilibrium (zero current) value, to have a given current density flowing through the electrode surface. This is an important realization: there can be no useful current flowing at the equilibrium potential (no products formed in electrolysis, and no power given off from a fuel cell); to have useful current, we have to have a finite overpotential. The overpotential always "opposes" the process: this forces us to use higher voltage (put in more electrical energy) in a process producing a chemical by electrolysis than that predicted by the equilibrium potentials, obtained from handbooks; and this makes less power available from batteries and fuel cells than one would expect from the equilibrium potentials. This is called "polarization," and already was mentioned in Chapters 4 and 5. While we have to live with polarization, as an unavoidable natural phenomenon, we can minimize it, as further examination of the equation will show later.

The other terms of the exponential are constants: F is the faraday number, which was already mentioned, R is the universal gas constant, and T is the temperature. The constant β is intimately connected with the actual physical electron transfer step and can be given realistic meaning only if the quantum-mechanical tunneling nature of the electron transfer is considered, something we are not planning to get involved with now. The pre-exponential constant i_0 is important enough to merit a separate section later.

A closer scrutiny of the last equation will reveal to the mathematically minded that the current is not zero at zero overpotential (the equilibrium potential). But we have already stated that there is no useful reaction possible at this potential. Is not this a contradiction? Well, we have looked so far only at half of the situation. While the above expression is true for the reaction as written, the reaction can also proceed in the opposite direction:

$$A \leftarrow B + e$$

and this reaction will take place at a rate of

$$i = i_0 \exp\left(- \beta \frac{F}{RT} \eta \right)$$

At any value of the solution–metal potential difference, or to talk in terms of measurable quantities, at any value of the overpotential η, both of these reactions proceed simultaneously, but usually at different rates, and the overall, net reaction rate is the difference of these two rates:

$$i = i_0 \left[\exp\left(1 - \beta \frac{F}{RT} \eta \right) - \exp\left(- \beta \frac{F}{RT} \eta \right) \right]$$

This important basic equation of modern electrochemistry is called the Butler–Volmer equation. Now, it is best to illustrate this graphically. Fig. A2.3 shows the forward, the backward, and the overall reaction rates as functions of the overpotential. When the overpotential is positive, the rate of the forward reaction (as it has been written above) is larger, giving a net forward rate (by definition, the current is then positive). When

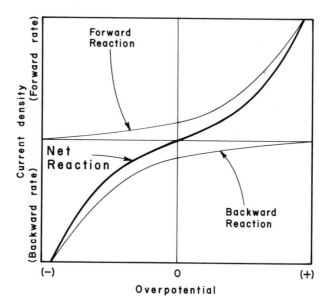

Fig. A2.3. The basic current density–overpotential relations.

η is negative, the backward reaction dominates the rate (the current is then, by definition, negative). At the equilibrium potential ($\eta = 0$), both rates are equal, and therefore there is no *net* current, no *net* reaction at $\eta = 0$.

A2.7. The Equilibrium Potential

We have been mentioning the equilibrium potential lately; some further examination of it is warranted. As we have discussed in Appendix I, when a metal is immersed into a solution, a potential difference will always exist between the two phases. If the electrode is not connected, through an outside circuit, to another electrode, and the net current is zero, this potential difference is the equilibrium one. While this potential difference is not measurable, we still can give it a relative value on our arbitrary scale (see Appendix I) as measured against our universal reference electrode, and it can be expressed by the classical Nernst equation, for our reaction, assuming that n electrons take part in it when it occurs once:

$$V = V^0 + \frac{RT}{nF} \ln \frac{C_B}{C_A}$$

The value of the potential depends on the concentrations of the reacting species (C_A and C_B) and the temperature (R, T, and F have the same physical meaning as in the current–potential equation above); V^0 is the standard value given in handbooks for unit concentrations.

The concentration dependence of this potential can be practically utilized for analytical purposes, as already discussed in Chapter 10. Some further effects of this concentration dependence will be examined in Appendix III.

A2.8. The Exchange Current Density and Electrocatalysis

Let us go back to our current–overpotential relations and to the neglected pre-exponential term. A meaning was already given to it tacitly: i_0 is the current density which is flowing in *both* directions at the equilibrium potential (no net current). It is the current density being "exchanged" between the forward and backward reactions, and hence the name: *exchange current density*. While this is a fairly straightforward physical meaning, there is more depth to it than that. What we have not mentioned up to now is the fact that the rate of a given reaction can be fast or slow depending on the electrode material used. For example, the exchange current density, and so the rate of the reaction, of the hydrogen evolution reaction ($2H^+ + 2e \rightleftarrows H_2$) can change almost ten orders of magnitude (by almost ten billion times) simply by changing from one electrode metal to another. This phenomenon is called *electrocatalysis*. Similarly to chemical catalysis, on first examination the catalyst does not seem to take part in the reaction (it does not appear in the overall reaction), but it can greatly influence its rate. Of course, the electrodes (and catalysts in general) do take part in the reaction. As we have already discussed, the reacting species must be in the interphase, in that region of the solution which is under the close influence of the metal phase. It should not be surprising, therefore, to find that the nature of the metal affects the rate of the electron transfer.

The importance of electrocatalysis cannot be overemphasized. When the possibility exists of increasing the current, at a given overpotential, by many orders of magnitude, the economic justification of research in electrocatalysis is obvious. We are not going to go into the theory of this subject; it is quite complex, and not well developed. Let it suffice to say that it is one of the newest fields of electrochemistry (about 10 years old), one in which considerable opportunities, and challenges, exist.

Some further, general considerations relating to this matter will be introduced here. If you go from Philadelphia to New York by plane, you will probably spend more time going from the New York airport to the center of town than in the flight between the cities. If you want to carry out an experiment in a laboratory, the experiment may take a few weeks, but getting the equipment often takes a year or more. These examples of rate-determining steps ("getting from the airport to the town") in processes ("going from Philadelphia to New York"). What we do always has a chain of sequences. We want to accomplish a certain act P, but before we reach it we have to go through the acts A, B, C, . . . until we get to P. Thus, to obtain our goal, the act P, we must carry out a series of associated acts, and one of them usually is a rate determiner, i.e., it is the bottleneck for all the others.

Let us exemplify this in electrochemistry, because knowing something of rate-determining steps is essential if we are going to understand the important part of electrochemistry called electrocatalysis (which, as we have seen, is what determines the price of electrochemically made goods).

Suppose that we consider electrochemical reactions which are a bit more complicated than those which we have been talking about above. We might consider a quite simple example, the overall reaction for the deposition of hydrogen. First, we start with the discharge of the proton:

$$H^+ + e \rightarrow H_{ads}$$

Then consider the combination of atoms to give us hydrogen molecules:

$$2H_{ads} \to H_{2\,gas}$$

Thus, the overall reaction is

$$2H^+ + 2e \to H_2$$

The rate-determining step, then, could be either the electron transfer or the combination of hydrogen atoms.

Suppose that the rate-determining step is the combination of hydrogen atoms. This means that the forward reaction, that of the protons being transferred from the electrolyte to the metal—with acceptance of electrons on the way—is faster than the combination of hydrogen atoms. In this case, of course, there would be a build-up of hydrogens on the surface, and these atoms would then start to go back again into the solution. Thus, there is a forward reaction rate of the transfer of protons to the electrode and a backward reaction rate of the transfer of hydrogen atoms back to protons, with injection of electrons into the metal. These rates are almost, but not quite, equal, and the small difference between them is equal to the rate of combination of hydrogen atoms to form molecules.

Thus, the rate of actually forming H_2 is controlled by the rate-determining step—the hydrogen combination in this example—and what would be happening before this would be a kind of pseudoequilibrium. The rate of the reaction of protons going from the solution onto the electrode is not quite equal to the rate of that going back again. But if the rate constant for the combination reaction is relatively *small*, then it is likely that the rate of protons going from the solution to the electrode and that of H atoms going back into the solution will be so much greater than the rate of combination of hydrogen atoms to molecules that we can consider the first reaction as "almost" in equilibrium.

Why should we bother about all this? Well, the point is that if we want to make a reaction go faster,* then we must

---

* Or have a lesser overpotential for the same rate, and hence have to use up less electric power to produce the same amount of a chemical.

have a faster rate constant—a higher exchange current density. This means, in turn, that we want to have a fast rate-determining step.

But we cannot evaluate, or determine, the rate-determining step until we have identified it—found out which of several possibilities it is in a reaction sequence. Let us first, therefore, realize that *all electrochemical reactions have one rate-determining step*, which is more important than any other step. It is this step that we have to work upon if we are going to understand and improve electrocatalysis. To try to increase electrocatalysis without studying the rate-determining step in the reaction is like trying to speed up a stalled journey without first finding out what is holding up the journey.

In electrochemistry the determination of reaction mechanisms is often a fairly complex affair. Further, it is not always possible to complete the determination because there are many aspects not yet understood. However, if we concentrate on just one aspect of the determination of mechanism—the determination of the rate-determining reaction step—then it is easier (it is less important to know the identity of the other reactions).

There are many methods for the determination of mechanism and we will mention only one here. Looking back at our basic current density—overpotential relation, we have to admit that we have written it in its simplest possible form; it relates, as it is, only to the simplest electrochemical reaction consisting of only one single step: an electron transfer (of course, that step must then be the rate-determining step). When the reaction is more complex (series of steps), more constants appear in the exponential terms, and these constants depend on the actual mechanism. The experimental determination of the exact nature of the exponential current—potential relation can sometimes be an aid in mechanism determination.

A2.9. Summary

We have dealt with a basic, perhaps surprising, and ecologically important fact: electricity can convert one form of matter to another and changes in matter can be converted directly to electricity. We broke down the rate of reaction at an interface into two currents, a forward current and a backward current. They both flow across interfaces all the time, though one can be much greater than the other. Furthermore, we have shown that there must be two electrodes in any electrochemical setup—like the necessity of having two hands for clapping—and both together are arranged to form a cell.

Then we gave a somewhat complicated looking equation: the relation between the potential difference in the double layer and the rate of an electrode reaction. This is the equation which underlies much of electrochemistry: the synthesis of organic compounds, the recovery of the metal from junked cars, the acceleration of a car running on a fuel cell, or the conversion of materials directly to electricity.

Just after that we gave a few more details, for example, the concepts of exchange current density and electrocatalysis.

The relevance of all these to the practical matters discussed earlier in the book must be made clear. It is necessary that we clean up our environment, but we have to be able to do it at an acceptable cost; it is necessary to have clean energy sources, but the price of the power must not be exorbitant. It is here, with respect to the cost of the processes, that the above-mentioned rate considerations become important, because the higher the rate constants or exchange current densities are, the less is the overpotential. But if this overpotential is small, so is the applied potential for a given current and hence the cost of electricity to produce a given weight of material.

APPENDIX III

Transport of Charges to and from an Interface

A3.1. Why Logistics Is Important in Electrochemical Science

We have talked so far about what happens when the ion is actually at the interface. What does this mean? It means that the ion is a few angstroms from the interface. This is the sort of distance across which electrons transfer. But we must ask: how do the ions get there? For example, if we are going to transfer electrons from an electrode to the ions, the ions have to be brought from the interior of the solution. If we are going to have ions which reach the electrode, give up or collect a charge, and remove themselves from the interface, they have to do so by some mechanism.

The mechanisms of ionic transport to and from the electrode are not mere academic questions, something for the mathematically oriented electrochemist. They have great importance to electrochemistry as a practical science. It is questions of transport which may limit, for example, the power density of electrochemical energy converters or the rate at which we can make electrode reactions take place. The rate at which electricity can be transferred across an interface—the basic electrochemical step—cannot *exceed* that at which we can transport charges to the interface. Logistics is, eventually, the master of transfer.

A3.2. A Simple View of Transport in an Electrode Reaction

Let us consider a cathodic electrode reaction and deal with our favorite example, the donation of electrons from a solid phase to protons in solution. Suppose that this is taking place at some reaction rate, i amperes per square centimeter. Then, suppose that we consider the rate at which the ions will come to the interface. Obviously, if we are dealing with a steady state, that is to say, the current is not changing with time, then there must be a rate of transport to the electrode equal to the rate at which the electrons transfer across the interphase to the protons.

We now make a simplification. Let us assume that there is only one method by which we can gain access to the electrode space, and that this method is *diffusion*. What we are saying is that in the steady state the rate of diffusion to the electrode is equal to the rate at which the electrons are transported from the electrode, across the double layer, to the H^+ on the other side (i.e., the solution side) of the double layer.

Now, we can refer to books which deal with diffusion and find out what the laws of diffusion are. One very simple, empirical law (and an old one) is that of Fick. According to this, the steady-state diffusion (the diffusion rate which does not change with time) is given by

$$\text{rate of transport} = -D\frac{dC}{dx}$$

This means that the diffusion rate is proportional to the *concentration gradient*. The minus sign means that the ions will go in the direction of decreasing concentration. The proportionality factor, D, is called the diffusion coefficient.

Now we consider a simplification due to the renowned classical electrochemist, Nernst. The concentration gradient near the electrode is linear, and we assume that it ends sharply (see Fig. A3.1A). The real situation is shown in Fig. A3.1B: the approximation is not bad. Looking at this figure, we can rewrite Fick's law as

$$\text{rate of transport to the electrode} = -D\,\frac{C_e - C_b}{\delta}$$

Thus, when the supply of ions demanded by the electron transfer rate at the interface is small, the difference between the concentration at the electrode, C_e, and the concentration in the bulk, C_b, of the protons will be negligible. But when the current density gets higher, there will be a large difference between C_e and C_b and we shall see that the diffusion rate will increase until it is equal to the current demanded by the electron transfer rate at the electrode.

But, if we look at the diffusion equation, we see that, as the value of i keeps on going up, the value of C_e keeps on going down, so that there will eventually come a time when there will be a value of $C_e = 0$ (Fig. A3.2).

With no ions present at the interface, the current density will become the so-called "limiting current." This is the maximum value of the current which an electrode can support, or give rise to, and no current, carried by the ions concerned, can be more than this. Its value is

$$i_L = nFD\,\frac{C_b}{\delta}^{*}$$

This is an example of logistics governing transport.

A3.3. Making the Limiting Current Bigger

If we are trying to remove poisonous materials from a liquid in a factory (e.g., cyanides), it may be important to make the limiting current as large as possible. We have just seen the equation which gives the limiting current. We cannot change

---

* The flux is multiplied by nF to give the current (*cf.* Section A2.5).

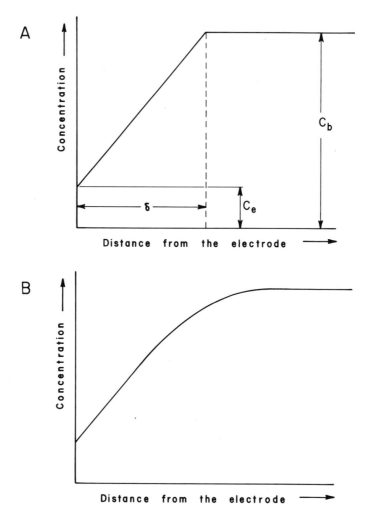

Fig. A3.1. (A) The Nernst approximation of the diffusion layer. (B) The true concentration–distance relation.

the diffusion coefficient in it—that is something that has to do
with the physical chemistry of solutions, and every ion or organic
substance will have its own particular diffusion coefficient.

   The concentration in the bulk of the solution is also something determined once and for all by the solution, and it seems

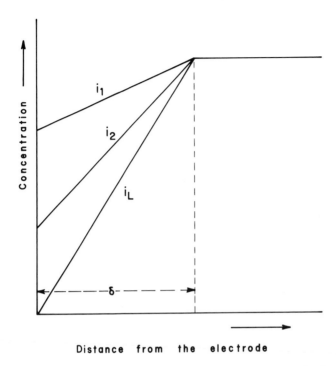

Fig. A3.2. The change of the concentration gradient (and surface concentration) with increasing current ($i_1 < i_2 < i_L$).

that the only thing which we might be able to change would be the value of δ, the "thickness of the diffusion layer." We made it clear in Section A3.2 that δ is a simplification. The linearized diagrams that we have given show that it is a matter of making a simple job out of a complicated one. Consider, then, what would happen if we stirred the solution with a paddle. Of course, there would be more material forced up through the electrode by the action of the paddle (we call this process forced convection). The point in the solution, near the electrode, at which the diminution of concentration will begin would now be nearer to the electrode than in an unstirred solution. This means that δ, the magnitude of the diffusion layer thickness, would be reduced, and this would make the limiting current larger (*cf.* the equation in Section A3.2).

In practice, by stirring the solution in some way (e.g., by imposition of ultrasonic waves upon the electrode), we can reduce the thickness of the diffusion layer by a factor of about 50, i.e., we can make the limiting current 50 times greater than the value we would get in an unstirred solution.*

Stirring is important and can be achieved in several ways. An increase of temperature also helps—for this increases the value of D. Correspondingly, boiling—greatly agitating the solution—greatly decreases δ and increases the i_L.

A3.4. Natural Convection

There is another phenomenon connected with transport which we must understand. Understanding it helps us to realize why the value of δ, the diffusion layer thickness, is constant, independent of time, after a little while. Thus, if we consider a supersimple system in which nothing happened after we switched on the current, except that electrons are transferred to protons, and protons diffuse from the bulk to the electrode, we should expect the value of δ gradually to grow. Thus, at the beginning (let us think of 0.1 sec), there would be very few of the protons used up, in the immediate vicinity of the electrode, so that the dimensions of the diffusion layer would be small. However, as the process of removal of protons and forming of the hydrogen went on, more protons would be used up. Therefore, the layer of solution near to the electrode in which the protons are being removed would get greater. The diffusion layer would get thicker. Eventually—quite hypothetically—the diffusion layer would extend from the electrode to the edge of the experimental apparatus. This sounds improbable and indeed intuition tells us that some other process would intervene. It is the effect of gravity. For, as we change the concentration near the electrode, the specific gravity of the layers changes. The layers nearest the electrode are the most denuded, and they tend to rise, the layers further out tend to fall (see Fig. A3.3).

This process is known as "natural (i.e., unforced by some-

---

* In unstirred solution δ is about 0.05 cm.

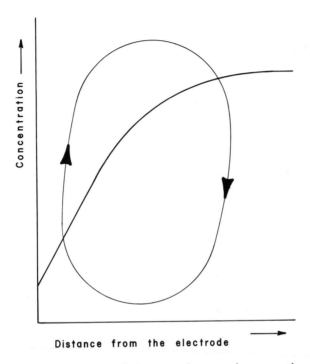

Distance from the electrode

Fig. A3.3. Origin of the natural convection near the electrode. The low-density (low concentration) solution rises, while the denser (more concentrated) solution sinks.

thing from outside) convection," and it results in a kind of "sealing off" of the diffusion layer. Only near to the electrode is anything simple happening, and further out, there is convection, stirring up the electrolyte (hence, making the concentration of ions in it uniform) without stirring being introduced artificially by some device. We now understand why, in practice, δ does not go on increasing until it reaches the edge of the experimental apparatus.

A3.5. Field-Dependent Transport

There is a third type of transport which we have to take into account, and that is one which involves transport to the electrode by means of the electric field which we impose on the solution.

This field is that which exists between the electrodes. It is a macrofield, nothing to do with the atomic scale, and we calculate it by measuring the voltage drop which exists between the two electrodes and dividing by the distance. This field exerts its effect upon the ions, accelerating the positively charged ions toward the negatively charged electrode, and the negatively charged ions toward the positively charged electrode. This mode of transport is not free—we pay for it since the voltage (IR) drop in the solution is included in the cell voltage.

A3.6. The Concentration Overpotential

We have left overpotential out of our discussions so far and we must hasten to bring it back again. To connect up overpotential with transport considerations, let us isolate the overpotential from considerations which have imbued our thinking in Appendix II, those of getting the electrons across the double layer at some required velocity, and suppose this part of things has been fixed.

For example, if we look at the Butler–Volmer equation (Section A2.6), then we can see that, if i_0 is sufficiently large, the overpotential arising from electron transfer is small. It is true that this need not be the case—it usually isn't—but it will make our considerations easier if we deal with a situation for which i_0 is assumed to be so large that (*cf.* the Butler–Volmer equation) the charge transfer overpotential is negligible.

Accordingly, there will be no overpotential at any current density. But remember how overpotential was defined: it is the difference between the actual electrode potential and the equilibrium potential, and in the consideration of the charge transfer overpotential, it was assumed that the equilibrium potential remains constant during the electrode process. If, however, the concentration of the reacting species changes at the electrode surface because of transport difficulties (slow diffusion), as described above, the equilibrium potential will change (the electrode "senses" only what is in its immediate vicinity, and not the bulk solution concentration). The potential shift can be calculated from the Nernst equation (Section A2.7), and is called the concentration overpotential:

$$\eta_c = \frac{RT}{nF} \ln \frac{C_b}{C_s}$$

or, using the relations given in Section A3.2 for the limiting current density to express the concentration ratio,

$$\eta_c = \frac{RT}{nF} \ln \left(1 - \frac{i}{i_L} \right)$$

It is easy to see, then, that the overpotential arising from concentration gradients will be small unless the current density is close to the limiting current. Hence, it is only when current densities approach more than, say, 10% of the limiting current

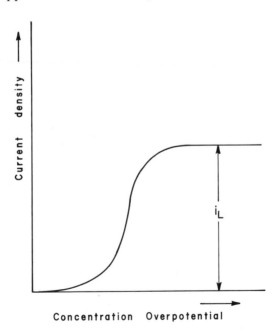

Fig. A3.4. The concentration overpotential–current density relation.

for the solution concerned, that concentration overpotential will be large enough for us to have to take account of it.

Let us consider what would happen during the increase of the current. In the diagram shown in Fig. A3.4 we see that the concentration overpotential increases rapidly once we get near the limiting current. Hence, once we do get near the end of the supply of ions, there is a big increase of overpotential, to such an extent that the potential is increased in vain, the current stays constant: the limiting current has been reached.

In a practical situation, both the charge transfer and concentration polarizations exist simultaneously. The decrease of polarization, as we emphasized in Appendix II, is a practical matter, when one is interested in the cost of the electrochemical processes. The same is, of course, also true for the concentration polarization. Introduction of forced convection (stirring, circulation, etc.) into the electrochemical cells and design features enhancing natural convection are important facets of electrochemical engineering.

Index